ACS SYMPOSIUM SERIES **414**

Carcinogenicity and Pesticides
Principles, Issues, and Relationships

Nancy N. Ragsdale, EDITOR
U.S. Department of Agriculture

Robert E. Menzer, EDITOR
University of Maryland

Developed from a symposium sponsored
by the Division of Agrochemicals
at the 196th National Meeting
of the American Chemical Society,
Los Angeles, California,
September 25–30, 1988

American Chemical Society, Washington, DC 1989

Library of Congress Cataloging-in-Publication Data

Carcinogenicity and pesticides.

(ACS symposium series, 0097–6156; 414)

"Developed from a symposium sponsored by the
Division of Agrochemicals at the 196th National Meeting
of the American Chemical Society, Los Angeles, California,
September 25–30, 1988."

Includes bibliographical references.

1. Pesticides—Carcinogenicity—Congresses.
2. Carcinogenesis—Congresses.

I. Ragsdale, Nancy N., 1938–   . II. Menzer, Robert E.,
1938–   . III. American Chemical Society. Division of
Agrochemicals. IV. American Chemical Society. Meeting
(196th: 1988: Los Angeles, Calif.) V. Series.

RC268.7.P47C37   1989   616.99′4071   89–18052
ISBN 0–8412–1703–3

The paper used in this publication meets the minimum requirements of American
National Standard for Information Sciences—Permanence of Paper for Printed Library
Materials, ANSI Z39.48–1984.

PRINTED IN THE UNITED STATES OF AMERICA

Foreword

The ACS SYMPOSIUM SERIES was founded in 1974 to provide a medium for publishing symposia quickly in book form. The format of the Series parallels that of the continuing ADVANCES IN CHEMISTRY SERIES except that, in order to save time, the papers are not typeset but are reproduced as they are submitted by the authors in camera-ready form. Papers are reviewed under the supervision of the Editors with the assistance of the Series Advisory Board and are selected to maintain the integrity of the symposia; however, verbatim reproductions of previously published papers are not accepted. Both reviews and reports of research are acceptable, because symposia may embrace both types of presentation.

Contents

Preface

XENOBIOTIC CARCINOPHOBIA, the fear of cancer resulting from foreign substances, has strongly influenced the public to demand regulatory action affecting pesticides that have been determined through established protocols to cause cancer. Many uncertainties exist, as discussed in this book, in the interpretation of data to predict whether a pesticide will induce cancer in humans. This is further complicated by the fact that carcinogenesis itself is not completely understood. Cancer is indeed a serious affliction of concern to all people, but a balance must be struck between the necessity to provide food and fiber for an expanding world population and our ability to avoid completely all health risks. Realistically, the production of food and fiber without pesticides is not possible in today's world. Thus, using our current state of knowledge, we must endeavor to establish a level of risk that will be acceptable in light of the benefits to be achieved.

This book is based on a symposium that considered the process of carcinogenesis itself, the external factors, including pesticides, which influence the process, and how society should perceive and react to pesticides with respect to carcinogenicity. A broad set of topics is covered, ranging from theoretical to practical considerations. A historical perspective is also presented.

We would like to express our appreciation to the Cooperative State Research Service of the U.S. Department of Agriculture for its interest and support of the symposium. We thank Elizabeth Weisberger, John Doull, and Christopher Wilkinson for their excellent advice in the planning phase.

NANCY N. RAGSDALE
U.S. Department of Agriculture
Washington, DC 20250

ROBERT E. MENZER
University of Maryland
College Park, MD 20742

Current address:
Environmental Coordinator's Office
University of Maryland
College Park, MD 20742

Current address:
Environmental Research Laboratory
U.S. Environmental Protection Agency
Gulf Breeze, FL 32561

June 28, 1989

Chapter 1

Pesticide Carcinogenicity

Introduction and Background

John Doull

Department of Pharmacology, Toxicology, and Therapeutics, University of Kansas Medical Center, Kansas City, KS 66103

Although the focus of this book is on pesticide carcinogenicity, pesticides can also produce other types of delayed toxicity such as reproductive effects, fetal damage, delayed neurologic manifestations, possible immunologic disorders and other adverse health effects. We also need to consider the relative impact of cancer and other delayed effects of pesticide exposure on human health when compared with the problem of acute pesticide poisoning. Epidemiologic and occupational medicine data clearly indicate that acute pesticide poisoning is the most frequently encountered and best clinically documented adverse health effect associated with pesticide exposure. Despite this, the major concern of the media, the public, and most other groups seems to be focused on the carcinogenic effects of pesticides. Oncogenicity studies are required for the registration of pesticides, and since it is these studies which provide the basis for our predictions regarding pesticide carcinogenicity, we should consider whether we may be over-interpreting these results and neglecting other toxicologic findings.

All of us are concerned about the relationship between pesticides and cancer. However, cancer is not the only adverse effect of pesticides. It is probably not even the most important adverse effect of pesticides. Furthermore, just because a pesticide causes cancer in rats or in mice does not necessarily mean that it will cause cancer in humans.

Development of Pesticide Regulation

The general approach for evaluating the potential adverse effects of pesticides in non-target species was established over four decades ago by Lehman, Fitzhugh and Nelson at the Food and Drug Administration (FDA) which was responsible at that time for establishing pesticide tolerances on food in the United States. The approach, which was similar to that used for drugs, involved a case-by-case comparison of the risk/benefit ratio for each pesticide. The evaluation of risk was based primarily on two types of toxicity studies in animals: acute studies, which included LD_{50} determinations coupled with a description of the symptoms and adverse effects, and chronic studies, which were multidose 1-2 year rodent studies. The intent of this approach was to identify all of the potential adverse effects that could be produced by either acute or chronic exposure to the pesticide and to determine the exposure dose of the pesticide required to produce all such effects. Subchronic (90 day) studies were utilized to establish the dose levels for the chronic studies and additional kinetic, metabolism and mechanism of action studies were carried out if needed to characterize the toxicity of the pesticide. Both the acute and the chronic data were then used to establish thresholds for the adverse effects of the pesticide, and in most cases, the tolerances and other regulatory positions were established by dividing the lowest threshold by an appropriate safety factor. This

0097–6156/89/0414–0001$06.00/0

approach, which was subsequently adopted as the acceptable daily intake (ADI) method for food additives, was and is used both in the U.S. and abroad for regulating pesticides as well as other classes of chemicals.

During the subsequent two decades preceding the creation of the U.S. Environmental Protection Agency (EPA) and the transfer of the major responsibility for regulating pesticides to EPA, there were several additions to the toxicity testing requirements for evaluating the hazard of pesticides. The discovery of organophosphate insecticide potentiation by Frawley at the FDA led to a requirement for testing of all new organophosphorus insecticides for such synergistic effects. A leghorn chicken test was also required for all new organophosphorus insecticides to detect any delayed neuromuscular effects of these agents. The thalidomide episode stimulated the development of teratology as a part of the toxicologic testing procedures. Growing concern over other types of fetotoxicity stimulated the development of multi-generation protocols for evaluating fertility, post-natal, and other reproductive effects of pesticides.

In more recent times methods for detecting and evaluating the genotoxic effects of pesticides were stimulated by the studies of Bruce Ames and his associates. Significant developments have occurred in the areas of neurotoxicology, behavioral toxicology, and immunotoxicology, but these have not yet resulted in validated protocols. Thus, they have not been incorporated into regulatory requirements for pesticide registration. However, the EPA is actively trying to develop guidelines for pharmacokinetics, neurotoxicology, behavioral and immunotoxicology testing even in the absence of validated protocols. Although these changes in toxicologic methodology provided a broader data base for predicting potential adverse effects in non-target species, the basic approach in pesticide regulation continued to rely on both the acute and the chronic data for hazard evaluation and the setting of pesticide tolerances.

The current approach to pesticide regulation differs from this earlier approach in two ways. First, it is focused on the chronic effects of pesticides rather than on the acute effects, and second, it is focused on a single chronic effect, namely cancer. Emergency room physicians who treat pesticide poisoning have pointed out that it is the acute effects of pesticides that are responsible for most of the actual morbidity and mortality associated with most types of pesticide poisoning. They argue that we should be more concerned about the acute effects which are "real-world" than about the cancer risks which are largely hypothetical. Further, the current protocol used for chronic studies on pesticides is based on the National Cancer Institute (NCI) oncogenicity bioassay rather than on the more holistic procedure which was previously required. The NCI protocol was designed to serve as a screening test for carcinogenicity. Consequently it is neither adequate nor appropriate as a bioassay for chronic toxic effects other than cancer.

Another major difference between the current approach and the previous approach for evaluating the adverse effects of pesticide exposure involves the way in which the toxicity data are interpreted. In the current system, the carcinogenicity data from the chronic rodent studies are extrapolated using mathematical models which provide a numerical estimate of the upper bound of the cancer risk, and these numbers (Q^* values) are then used for a variety of regulatory purposes. In essence, this approach substitutes mathematical guidelines for the scientific judgement that was the key element in the earlier approach.

Since the basic purpose of conducting toxicity studies is to provide an accurate prediction of the potential adverse effects of a chemical in non-target, as well as in the test species, we need to ask whether the methodology that we use both for testing and for interpreting the test results is giving us answers that protect both the environment and public health. Based on the arguments presented in the preceding discussion, it is likely that our predictions regarding the adverse effects of pesticides on humans would be improved if they included a greater emphasis on the acute effects and on the non-cancer chronic effects of pesticide exposure. Protocols for detecting both acute and chronic neurotoxicity and the behavioral effects of pesticides, including specific tests for neuronal dysfunction, should be part of the requirements for pesticide registration. Recent workshops devoted to these areas (LSRO [Life Science Research Office] report to FDA on "Predicting Neurotoxicity and Behavioral Dysfunction from Preclinical

Toxicologic Data" and the "Workshop on the Effects of Pesticides on Human Health" which was organized by the Task Force on Environmental Cancer, Heart and Lung Disease) have recommended methods and test batteries that would detect sensory, motor, autonomic, cognitive and behavioral dysfunction.

Endocrine and immunologic related adverse effects of pesticides are another example of an area where we need toxicologic testing methods that will detect both acute and chronic toxicity and provide a reliable basis for predicting human effects. Before we add new protocols to the current methodology, however, we need to evaluate the answers provided by our current methodology.

Another area that would be of direct benefit to the medical personnel responsible for treating pesticide exposure in humans is the development of antidotal and symptomatic treatment for the various classes of pesticides. Although we have more specific antidotes available to treat pesticide poisoning than for any other single class of acute poisons, the number of antidotes is small for non-insecticidal pesticides and these cases must be treated symptomatically. The current acute toxicity protocols do not currently provide the kind of clinical information that is needed in such cases.

Development of Data for Evaluation of Effects

The two scientific disciplines which contribute most directly to the evaluation of cancer and all other human health hazards that may result from exposure to pesticides and all other toxic chemicals are **epidemiology** and **toxicology**. When these disciplines function in this "watch-dog" role, both have the same goal: to gather sufficient information about the potential adverse effects of exposure to the chemical so that we can reliably predict both the type of adverse effects that may occur and the exposure conditions that are likely to produce each of these anticipated adverse effects. Although the primary difference between these two disciplines is the research subjects (epidemiologists use humans whereas toxicologists use rodents), there are also major differences in how the problem is approached, the techniques used to generate the data, and how the data are used to predict the hazards. Toxicology and epidemiology start and end at the same place, but get there by different paths.

It should be pointed out here that there is no disagreement between epidemiologists and toxicologists as to the priority of human data over rodent data. Toxicologists recognize that in those situations where sufficient human data exist, there is no need to do any type of toxicity studies, unless we are predicting for environmental or non-human effects. The main reason why toxicology rather than epidemiology serves as the basis for most regulatory decisions involving cancer and other adverse health effects is that we do **not** have sufficient human data on most of the chemicals for which predictions are needed. In addition, there are ethical constraints on the use of humans as research subjects, particularly with chemicals which are not drugs, that often preclude the type of epidemiologic studies we need to make reliable predictions for risk in man. Since it is obvious that we will need to continue to rely on the toxicology data base for making hazard or risk predictions, it is important that, in addition to searching for new methods for obtaining and interpreting the toxicology data, we also ask whether our current approach is giving us correct answers.

Laboratory Animal Tumor Data Extrapolation

One approach to the question of whether risk assessment methods based on the results of oncogenicity studies in rodents are giving us the right answers is to compare the predictions from the epidemiologic studies with those from the toxicology data base. When we do this for those agents for which we have predictions from both sources, there are only a few cases of discrepancies. We do not have animal data demonstrating that asbestos causes mesothelioma or lung tumors, and there is only limited evidence that arsenic causes cancer in rodents, although the human evidence for these chemicals is clearly sufficient. More recently, epidemiologists have concluded that the herbicide 2,4-D causes non-Hodgkins' lymphoma in exposed farmers (1) and that ethanol ingestion is associated with cancer of the esophagus, liver and possibly breast; but neither of these chemicals produces cancer in rodents. Although it has been argued that the epidemiologic evidence in these cases is weak, since it is based on an association rather than on a cause/effect link, these findings are of particular concern to toxicologists. They are not like asbestos where our failure to produce lung tumors can be attributed to the lack of a good animal model and inadequate testing. We now have 18 negative rodent studies for ethanol. The 2,4-D oncogenicity study was considered by the EPA FIFRA (Federal Insecticide, Fungicide, and Rodenticide Act) Science Advisory Panel to be an excellent study. Neither 2,4-D nor ethanol are mutagenic. Acetaldehyde, which is the primary metabolite of ethanol in both rodents and man, is also not mutagenic, although it did produce lung tumors in an inhalation study in rats. While it is reassuring to find that in most cases the toxicologic data supports the epidemiologic predictions, it is also clear that the exceptions must be investigated and resolved in order to maintain the credibility of our animal tests as reliable predictors of cancer and other adverse effects of chemicals in man.

We also need to look at the exceptions in the reverse situation in which the toxicology data are positive but the epidemiology is negative. Such exceptions can occur if the test species is more susceptible to the tumorigenic effect of a chemical than the target species. It has been suggested that the liver tumors produced by the halogenated hydrocarbons are an example of this situation. Many of the pesticides and solvents in this group, such as chlordane, heptachlor, trichlorethylene, methylene chloride, etc., produce liver tumors in rodents, but epidemiologic studies generally do not support these findings. It can be argued, of course, that the epidemiologic studies are inadequate, and that is certainly true for some of these agents, but it can also be argued that in these cases the rodent studies are over-predicting the cancer risk for man.

The B6C3F1 mouse, which is widely used as one of our rodent test species, does have a high incidence of liver tumors. It has been suggested by IARC that when positive results are only found in mice and not in other species, such results should not be used as a basis for classification as a probable carcinogen. Similar arguments have been made regarding the increased incidence of thyroid tumors in rats exposed to goitrogens, such as the sulfonamides. In this case the mechanism by which the thyroid tumors are produced appears to have a clear threshold (2). It has been recommended by an EPA Working Group that such agents be regulated on the basis of a no-effect level (NOEL) approach rather than with the conventional cancer models.

Another example of a case where our rodent tests appear to be giving us the wrong answer, or at least over-predicting the human risk, is the production of kidney tumors in the male rat by gasoline, paradichlorobenzene, limonene and tetrachloroethylene. The mechanism by which these agents cause kidney tumors in male rats appears to involve conjugation with an alpha-2u-globulin and the formation of hyaline droplets in the P2 segment of tubule. Since these droplets are not found in female rats, or other species, including humans, it is argued that male rat kidney tumors produced through this mechanism are not predictive for human carcinogenesis. On the basis of this argument and the lack of positive genotoxicity evidence for paradichlorobenzene, the Halogenated Organics Subcommittee of the EPA Science Advisory Board has recommended that the classification for this agent be downgraded from B2 to C in the new drinking water regulations for paradichlorobenzene.

During the past few months, the EPA has held workshops on the predictive value of mouse liver tumors, rat kidney tumors, thyroid tumors induced by various goitrogens, and the use of pharmacokinetic parameters in risk assessment. Some of the recommendations from these workshops are likely to be incorporated in the new Cancer Assessment Guidelines which the EPA is developing. One of the most common complaints about the use of mathematical models in risk assessment is that these models ignore the relevant biology. Partly in response to this criticism, a U.S. National Academy of Sciences/ National Research Council committee has issued a report which presents methods for incorporating pharmacokinetic data in risk assessment. When the recommendations of this committee were applied to the risk assessment of methylene chloride, the predicted carcinogenic risk was reduced by a factor of 7.

An approach that is receiving considerable current attention is to identify and quantify all of the uncertainties that exist in the mathematical modelling process so that we can design and carry out the research needed to substitute hard data for these uncertainties. These procedures should improve both the relevancy and the accuracy of the model-based approach, but we must also consider the more basic issue, which is whether there is sufficient scientific or biologic justification for extrapolating the results of our high dose rodent studies to the low doses needed for prediction.

Future Developments

As was pointed out previously, the current protocol for conducting chronic toxicity studies is based on the NCI oncogenicity bioassay which was a screening test designed primarily to answer two qualitative questions: is the chemical oncogenic and what kind of tumors are produced? When this program was transferred to the National Toxicology Program, some additional features were added to make it more like the conventional chronic toxicity study, and an effort was made to obtain an indication of the dose-response relationships. Basically, however, it is still a screening test for oncogenesis and the manipulation of the data with sophisticated mathematical models which generate finite risk estimates does not alter the quality of the input. The idea that these models somehow generate quantitative data which are valid far beyond the range of the actual data is an illusion. In any event, it is an idea that needs to be rigorously tested. At the same time, we need to re-examine the protocols that we use to generate the cancer data in animals to determine whether we can develop procedures that will provide quantitative information that can either be used on a stand-alone basis or as input for the extrapolation models.

Many of these issues are being considered by various scientific and regulatory groups, both in the United States and abroad. Some of the issues may be addressed in legislative proposals to revise the laws under which pesticides are regulated (HR 4737, HR 4739, the Pesticide Reform Bill, etc.). In this process of seeking remedies for actual or perceived deficiencies in the scientific methodology used to identify and quantify the hazards of pesticide exposure, we need to remember that the major problem is not methodological, but is, rather, the lack of data on which to base the prediction.

The real reason we have a problem is that we simply do not have enough data. But also we need to keep in mind the two points: the fact that a chemical produces cancer in rodents does not prove that it is going to be a carcinogen in people, and there are other effects caused by pesticides that are probably more important than cancer.

Literature Cited

1. Hoar, S. K.; Blair, A.; Holmes, F. F.; Boysen, C. D.; Robel, R. J.; Hoover, R.; Fraumeni, J. F., Jr. J. Am. Med. Assoc. 1986, 256, 1141-7; erratum: 256, 3351.
2. Paynter, O. E.; Burin, G. J.; Jaeger, R. B.; Gregario, C. A. Regul. Toxicol. Pharmacol. 1988, 8, 102-19.

RECEIVED September 1, 1989

Chapter 2

Pesticide Regulation Related to Carcinogenicity

James C. Lamb IV

Jellinek, Schwartz, Connolly & Freshman, Inc., 1015 15th Street, NW, Suite 500, Washington, DC 20005

Pesticides are registered and canceled on the basis of risk-benefit balancing by EPA under FIFRA. New products are only registered after review of complete toxicology, residue chemistry, and environmental fate data that are developed by the registrant. Many old products must be reregistered because they were initially registered without full toxicity testing. At its present pace, reregistration will not be completed and many old tolerances will not be reassessed until well into the 21st century. EPA establishes tolerances for pesticides on food. Carcinogenic pesticides may be registered and tolerances may be set for the raw agricultural commodity if the levels are safe. However, if the concentration of a pesticide on food increases when it is processed, a food additive tolerance is required. The Delaney clause of the Federal Food Drug and Cosmetic Act (FFDCA) does not allow a food additive tolerance to be set if the compound has been shown to induce cancer in man or animal. A narrow interpretation of the Delaney clause has prevented registration of new products, even when EPA believes the compound is safe and the upper-bound risks are estimated to be much less than one in a million. The EPA's Cancer Risk Assessment Guidelines are very important to the determination of whether or not a pesticide will be granted a food additive tolerance. The regulation of pesticide residues in food, and other issues such as groundwater contamination, consumer or worker exposure, and Agency resources, will profoundly affect the regulation of carcinogenic pesticides.

0097–6156/89/0414–0006$07.75/0
© 1989 American Chemical Society

The U.S. Environmental Protection Agency (EPA) is responsible for administering the Federal Insecticide, Fungicide and Rodenticide Act (FIFRA), pesticide law. FIFRA affects the use of many economically important products including fungicides, herbicides and insecticides, disinfectants and sterilizers, and rodenticides. The key to pesticide regulation is content of the pesticide label and the claims made by the pesticide registrant.

These products are considered valuable to the production of our food supply, for the control of disease, and for our general comfort. They also, by their nature as poisons to some living organisms, pose certain risks if not properly controlled or used.

Pesticide Regulation

The regulation of pesticides controls the distribution of pesticides through the registration of specific uses. New products are subjected to intensive review before they receive EPA registration. The standards for registration are extensive, and the safety of new products is rarely brought into question. The FIFRA provides EPA tremendous authority to require data prior to registration, and the Agency attempts to answer all significant questions relating to health and safety before a product is allowed to be marketed (see section on Data Requirements below). Even after registration, however, EPA can require the registrant to generate and submit additional data to support the pesticide registration.

Reregistration

Many products were registered by either USDA or EPA before modern toxicology and residue chemistry methods were developed. Therefore, many of the older pesticides are registered with a less-than-complete database. The process of reregistration is intended to bring older pesticides up to modern standards. Once completed, the process should establish confidence in the safety of these beneficial agents, but, until that process is completed, complaints will be aired that the pesticides are inadequately tested or insufficiently regulated (e.g., Mott and Snyder, 1988).

Special Review

Circumstances arise when new data, or reevaluation of old data, lead the Agency to conclude that an already registered pesticide may pose unreasonable risks of adverse effects based on toxicity and exposure data. The Agency can initiate a Special Review if any of the following criteria are met: acute toxicity in humans or domestic animals, potential adverse chronic effects, hazards to nontarget organisms, hazards to threatened or endangered species, or other adverse effects not anticipated by the Agency (40 CFR 154.7).

The Agency proceeds by announcing its decision to initiate a Special Review and describing the concerns that will be

reviewed in a Position Document 1 (PD-1). Of the chemicals
under review as of September 1987, most were being reviewed
because of their potential oncogenicity (Table I). Many of the
Special Reviews that have been completed also concern
oncogenicity as at least one of the criteria triggering review
(Table II).

If the Agency decides that the Special Review should not
proceed, it may publish a Position Document 2 (PD-2). These
may be returned to the registration standard process to finish
filling data gaps (Table III). This was done in the case of
2,4-dichlorophenoxy acetic acid (2,4-D) due to EPA concerns
regarding potential oncogenicity, but the Special Review was
stopped when the Agency concluded that the data were not
sufficient to demonstrate significant risk to humans (1). The
data showed that 2,4-D caused an equivocal increase in the
incidence of brain tumors in rats. Mixed results from human
studies were finally judged as insufficient. 2,4-D was
designated a group D carcinogen (insufficient data) by the EPA,
and the Special Review was concluded without a detailed
regulatory proposal by EPA.

When the Agency proceeds with the Special Review, it
conducts an extensive review of the risks and the benefits of
the pesticide. A proposed regulatory decision is then
developed and published in the Federal Register as a "Draft
Proposed Notice of Intent to Cancel," or Position Document 3
(PD-3). The Agency eventually publishes a final document,
which is a "Notice of Intent to Cancel" or Position Document 4
(PD-4).

The Special Review process is concluded if the registrant
accepts the conditions of the PD-4 and modifies the label
accordingly. This modification may include anything from minor
label changes to voluntary cancellation of some or all uses of
a pesticide. Examples of voluntary cancellations are given in
Table IV. If the registrant disagrees with the PD-4, it may
request a cancellation hearing and appear before an
administrative law judge to argue the case. At the end of the
Special Review process, the cancellation order is published by
the Agency and the label changes are enforced by the courts if
necessary.

Suspension and Cancellation

The use of the Special Review process is optional, and EPA can
decide to immediately issue a cancellation order or, in extreme
cases, an emergency suspension order. If EPA is confident that
additional facts would not change its ultimate decision and it
wants to expedite the regulatory action, it may publish the
notice of intent to cancel (NOIC) or notice of intent to
suspend (NOIS) or an emergency suspension order without the
steps of the Special Review process. The Agency also can
combine various steps of the Special Review process to expedite
the regulatory action. An emergency suspension, based on the
risks of developmental toxicity, was declared in the case of
dinoseb (2). Pesticides suspended, in part, for oncogenicity
concerns include 1,2-dibromo-3-chloropropane (DBCP) (3-6) and

TABLE I. Special Reviews Pending

Chemical	40 CFR 162.11 Criteria Possibly Met or Exceeded
Aldicarb	Acute Toxicity
Amitrole	Oncogenicity
Captafol	Oncogenicity Wildlife Effects (acute and chronic)
Captan	Oncogenicity Mutagenicity Other Chronic Effects
Carbofuran	Wildlife Effects
Chlordimeform	Oncogenicity
1,3-Dichloropropene	Oncogenicity
Daminozide	Oncogenicity
Dichlorvos (DDVP)	Oncogenicity Liver Toxicity
Ethylene Dithiocarbamates (EBDCs)	Oncogenicity Thyroid Effects Teratogenicity
Ethylene Oxide EtO	Oncogenicity Mutagenicity Testicular Effects
TPTH	Teratogenicity
Tributyltins	Acute and Chronic Toxicity to Non-target and Aquatic Organisms
2,4,5-Trichlorophenol (TCP) Fetotoxicity	Oncogenicity

SOURCE: Data are from ref. 17.

TABLE II. Special Reviews Completed

Chemical	40 CFR 162.11 Criteria Possibly Met or Exceeded
Alachlor	Oncogenicity
Amitraz (Baam)	Oncogenicity
Benomyl	Reduction in Non-target organisms Mutagenicity Teratogenicity Reproductive Effects Hazard to Wildlife
Cadmium	Oncogenicity Mutagenicity Teratogenicity Fetotoxicity
Carbon Tetrachloride	Oncogenicity Toxic Effects on Liver and Kidney
Chlorobenzilate	Oncogenicity Testicular Effects
Coal Tar and Creosote (non-wood use)	Oncogenicity Mutagenicity
Creosote (wood use)	Oncogenicity Mutagenicity
Cyanazine	Teratogenicity
DBCP (1,2-dibromo-3-chloropropane)	Oncogenicity Reproductive Effects Mutagenicity
Diallate	Oncogenicity Mutagenicity
Diazinon	Avian Hazard
Dicofol	Ecological Effects

Continued on next page

TABLE II. (cont'd)

Chemical	40 CRF 162.11 Criteria Possibly Met or Exceeded
Dimethoate	Oncogenicity Mutagenicity Fetotoxicity Reproductive Effects
Dinocap	Teratogenicity
Dinoseb	Teratogenicity Reproductive Effects Acute Effects
EBDC's (Ethylenebis- dithiocarbamates, maneb, mancozeb, metiram, nabam, zineb, amobam)	Oncogenicity Teratogenicity Hazard to Wildlife
Endrin	Oncogenicity Teratogenicity Reduction in Endangered Species and Non-target Species
EPN (Ethyl p-nitro phenyl thionoben- zenephosphorate	Neurotoxicity Hazard to Aquatic Organisms
Ethalfluralin	Oncogenicity
Ethylene Dibromide	Oncogenicity Mutagenicity Reproductive Effects
Goal	Oncogenicity
Inorganic Arsenicals (wood use)	Oncogenicity Mutagenicity Teratogenicity
Lindane	Oncogenicity Teratogenicity Reproductive Effects Acute Toxicity

Continued on next page

TABLE II. (cont'd)

Chemical	40 CFR 162.11 Criteria Possibly Met or Exceeded
	Other Chronic Effects
Linuron	Oncogenicity
PCNB (Pentachloro- nitrobenzene)	Oncogenicity
Pentachlorophenol and Derivatives (wood use)	Oncogenicity Fetotoxicity Teratogenicity
Pentachlorophenol (non-wood use)	Oncogenicity Fetotoxicity Teratogenicity
Pronamide	Oncogenicity
Sodium Fluoro Acetate (1080) No Antidote	Reduction in Non-target and Endangered Species
Strychnine/ Strychnine Sulfate	Reduction in Non-target and Endangered Species
2,4,5-T/Silvex	Oncogenicity Teratogenicity Fetotoxicity
Thiophanate Methyl	Mutagenicity Reduction in Non-target Species (Rebutted)
Toxaphene	Oncogenicity Population Reduction in Non-target Animal Species Organisms Acute Toxicity to Aquatic Organisms Chronic Effects to Wildlife
Trifluralin (Treflan)	Oncogenicity Mutagenicity

SOURCE: Data are from ref. 17.

TABLE III. Final Determination on Pre-Special Reviews
Returned to the Registration Process

Chemical	40 CFR 162.11 Criteria Possibly Met or Exceeded
Cacodylic Acid and Salts	Oncogenicity Mutagenicity Teratogenicity
Carbaryl	Oncogenicity Mutagenicity Teratogenicity
Chloroform (Trichloromethane)	Oncogenicity
Dichlorvos (DDVP Mutagenicity 2,2-dichlorovinyl dimethyl phosphate) Neurotoxicity	Oncogenicity Reproductive Effects Fetotoxicity
Diflubenzuron (Dimilin)	Hazard to Wildlife
Maleic Hydrazide MH	Oncogenicity Mutagenicity Reproductive Effects
Methanearsonates (includes Amine Methanearsonate, Calcium Acid Methanearsonate, Monoammonium Methanearsonate (MAMA), Monosodium Methane- arsonate (MSMA), Disodium Methane- arsonate (DSMA))	Oncogenicity Mutagenicity
Naled	Mutagenicity Fetotoxicity Reproductive Effects

Continued on next page

Table III. (cont'd)

40 CFR 162.11 Criteria Chemical	Possibly Met or Exceeded
OBPA (10,10'-Oxybisphenoxarsine) (some uses)	Arsenical Compound suspected: Oncogenicity Mutagenicity Teratogenicity
Paraquat	Emergency Treatment Chronic Effects
Piperonyl Butoxide	Oncogenicity
Rotenone	Oncogenicity Mutagenicity Teratogenicity Reproductive Effects Chronic Toxicity Significant Wildlife Populations Reductions Acute Toxicity to Aquatic Wildlife
1081 (Fluoro-acetamide)	Acute Toxicity to Mammalian and Avian Species Reduction in Endangered and Non-target Species Acute Toxicity without Antidote
Terbutryn	Oncogenicity Mutagenicity Reproductive Effects
Triallate	Oncogenicity Mutagenicity
S,S,S-Tributyl Phosphorothrithioate	Neurotoxicity
Tributyl Phophoro-Trithioate	Neurotoxicity
Trichlorfon	Oncogenicity Mutagenicity Teratogenicity Fetotoxicity Reproductive Effects

SOURCE: Data are from ref. 17.

TABLE IV. Final Actions--Voluntary Cancellations

Chemical	40 CFR 162.11 Criteria Possibly Met or Exceeded
Acrylonitrile (three products)	Oncogenicity Teratogenicity Neurotoxicity
Arsenic Trioxide	Oncogenicity Mutagenicity Teratogenicity
Benzene Oncogenicity (all products)	Mutagenicity and Blood Disorders
BHC	Oncogenicity
Captofol	Oncogenicity
Carbon Tetrachloride	Oncogenicity Toxic Effects on Liver and Kidneys
Chloranil	Oncogenicity
Chlordecone Kepone (Products of six formulators)	Oncogenicity
Copper Acetoarsenite	Oncogenicity Mutagenicity Teratogenicity
Copper Arsenate (Basic)	Oncogenicity
Endrin	Reduction in Endangered Species and Non-target Species
EPN	Neurotoxicity Hazard to Aquatic Organisms
Erbon	Oncogenicity Teratogenicity Fetotoxicity

Continued on next page

TABLE IV. (cont'd)

Chemicals	40 CFR 162.11 Criteria Possibly Met or Exceeded
Isocyanurates (some products)	Kidney effects
Monuron Oncogenicity (some products)	
Nitrofen (TOK)	Teratogenicity Mutagenicity Oncogenicity
OMPA (Octamethylpyro- phosphoramide)	Oncogenicity
Pentachlorophenol	Oncogenicity Fetotoxicity Teratogenicity
Perthane (many products)	Oncogenicity
Phenarsazine Chloride	None
Ronnel (many products)	Oncogenicity
Safrole	Oncogenicity Mutagenicity
Sodium Arsenite (two products)	Oncogenicity Mutagenicity Teratogenicity
Strobane	Oncogenicity
Trysben	Oncogenicity

SOURCE: Data are from ref. 17.

ethylene dibromide (EDB) (7). In both cases, the oncogenicity concerns were derived from animal studies. Other examples are provided in Table V.

Another pesticide that was suspended because of oncogenicity was chlordane (8-10). Chlordane was used on a number of crops, and the original EPA action was publication of a NOIC (this action preceded the development of the Special Review procedures). The cancellation hearing, however, took so long that the Administrator decided that suspending the uses was necessary while the cancellation hearing was in progress. The crop uses of chlordane were eventually voluntarily cancelled so that the action was ended without concluding the cancellation hearing. The crop uses of chlordane were cancelled, but some other uses, including the termiticide uses, remained.

The oncogenicity concerns with chlordane persisted until Velsicol and EPA negotiated a voluntary cancellation of most of the termiticide uses in 1987 and avoided a cancellation hearing. The cancellation hearing was expected to be hard fought, as the evidence of carcinogenicity was based on mouse liver tumors, which provided evidence of a very different weight in 1987 than in 1979.

Pesticide Data Requirements

Pesticide registration is a very data-intensive process. EPA has more data on pesticide toxicity than on any other type of chemical and, even though there are significant data gaps for many pesticides, the database on pesticide toxicity is quite large. The registrant of a new food-use pesticide is expected to provide a complete toxicology, residue chemistry, and environmental fate data package. A new product will not be registered without a complete data package. A pesticide used on food crops will not only need a registration under FIFRA, but also will need one or more tolerances for each crop under the Federal Food Drug and Cosmetic Act (FFDCA).

The registration of a new product requires the development of data costing millions of dollars prior to submitting the registration application to EPA. This means that bringing a new product onto the market is very difficult and time-consuming. By contrast, many older pesticides still in wide use today did not have a complete toxicology, residue chemistry, or environmental fate database when they were registered by USDA. Even if test data are available, they were probably generated by now-obsolete methods. Therefore, many old products must be reregistered. Reregistration involves bringing the database for older pesticide products up to modern standards. The data are all developed by, and at the expense of, the registrant. The cost of reregistration alone has led to the loss of various pesticide products of unknown toxicity. Other products are lost from the market as unfavorable data are developed. But in the end we will know more about the products that remain and should be better able to manage them safely.

EPA is obligated under the FIFRA to publish the kinds of information required to register a pesticide (FIFRA Sec. 3).

TABLE V. Notice of Intent to Cancel/Suspend Issued

Chemical	40 CFR 162.11 Criteria Possibly Met or Exceeded
Aldrin	Carcinogenicity Bio-accumulation Hazard to Wildlife and other Chronic Effects
Chlordane/Heptachlor	Carcinogenicity Reductions in Non-target and Endangered Species
Chlordecone (Kepone)	Oncogenicity
DBCP	Oncogenicity Mutagenicity Reproductive Effects
DDD (TDE)	Carcinogenicity Bio-accumulation Hazard to Wildlife and other Chronic Effects
DDT	Carcinogenicity Bio-accumulation Hazard to Wildlife and other Chronic Effects
Dieldrin	Carcinogenicity Bio-accumulation Hazard to Wildlife and other Chronic Effects
Dinitramine	Oncogenicity
EPN	Hazard to aquatic organisms
Mirex	Carcinogenicity Bio-accumulation Hazard to Wildlife and other Chronic Effects

Continued on next page

TABLE V. (cont'd)

Chemical	40 CFR 162.11 Criteria Possibly Met or Exceeded
Ronnel	Oncogenicity
2,4,5-T/Silvex	Oncogenicity Teratogenicity Fetotoxicity
Toxaphene	Oncogenicity Hazard to Wildlife and other Chronic Effects Reduction in Non-target Organisms

SOURCE: Data are from ref. 17.

The Agency has developed lists of typical test requirements under 40 CFR 158 (Part 158) including testing in the areas of product chemistry; residue chemistry; environmental fate; toxicology; reentry protection; aerial drift evaluation; toxicity to wildlife, aquatic organisms, and nontarget insects; plant protection; and product performance. Testing for carcinogenic potential is only one of many areas of concern to EPA. Specific tests in each of the categories may be required or waived for particular products. The Agency has reserved a great deal of discretion in the determination of which tests may be required for which products, but Part 158 serves to describe the general minimum or typical test battery.

Carcinogenicity Testing

Oncogenicity (or carcinogenicity) testing is one of the specific testing areas under Subdivision F of the Pesticide Assessment Guidelines: Hazard Evaluation for Humans and Domestic Animals, which describe typical details of the experiments that are required. The testing guidelines are more specific than Part 158. Part 158 helps the registrant know when a test will be required; the testing guidelines help it know how to conduct the test. Part 158 indicates that an oncogenicity testing is required for pesticides that will be used on food crops. Oncogenicity testing is conditionally required for nonfood uses if the use is likely to result in repeated human exposures over a significant portion of the human lifespan, or if the use requires a food tolerance or a food additive regulation. The testing requirement for oncogenicity is satisfied by testing two species, preferably

the rat and the mouse. The details of the typical protocol are
provided in the testing guidelines.

To assist the registrants in conducting acceptable studies,
EPA has published guidance documents on how to select a top
dose level that is the maximum tolerated dose (MTD) for cancer
studies (11). The guidance leads to establishing very high top
dose levels by setting rigid criteria for toxicity, which must
be observed for a study to be accepted by the EPA Office of
Pesticide Programs. The top dose level must be just below a
level that would result in life-threatening toxicity. Body
weight decrements of 10 to 15% are required to satisfy the
Agency that the MTD has been approached. An upper limit of 1
g/kg body weight/day has also been established for most
pesticides (2% or 20,000 ppm in the diet for rats, 0.7% or
7,000 ppm in the diet for mice).

Chronic toxicity and oncogenicity tests have been required
on all older products that are used on food crops. Failure to
provide the data on a schedule established by EPA will result
in the suspension of the registration of a product, until the
data are submitted. In some cases, the older products have
been shown to cause cancer in laboratory animals. Beyond the
information generated through the testing required under FIFRA,
very little information exists about the carcinogenicity of
pesticides. Almost no reliable information on humans exists;
therefore, human cancer risks are estimated from the animal
data, and pesticides are regulated based on these data. As
more chemicals are tested for carcinogenicity, as toxicity test
methods become more sensitive, and as analytical chemistry is
able to identify more products at lower levels, EPA will be
under even greater pressure: pressure from the public to
protect health, pressure from farmers to preserve agricultural
chemicals, and pressure from the agricultural chemical industry
not to overregulate.

Assessment of Cancer Risks for Pesticides

As mentioned above, EPA has required carcinogenicity (and other
toxicity) testing on all the food-use pesticides. Any new
food-use pesticide and many new non-food use pesticides must be
tested for carcinogenicity prior to registration by EPA. EPA
is currently requiring additional carcinogenicity testing on
nearly all of the products that begin reregistration. This
provides the Agency with a tremendous database on the carcinogenicity of these pesticides in rats and mice. The ultimate
goal of EPA is to determine the potential human cancer risks
associated with these products. The qualitative risks are
evaluated according to the Agency's risk assessment guidelines
(12).

EPA Cancer Risk Assessment Guidelines

The guidelines are written to provide guidance on Agency policy
for assessing cancer risk. They draw heavily upon the National
Research Council (13) approach to assessing risk by dividing
the risk assessment process into four activities: Hazard

Identification, Dose-Response Assessment, Exposure Assessment, and Risk Characterization. The guidelines lay out the current process in detail and stress the importance of scientific judgment in the process and that the classification system was not intended to be applied in a mechanical manner. These guidelines are being reviewed and revised by EPA at this time and many of the perceived problems may be addressed.

The current guidelines outline a scheme that classifies the weight of the evidence of carcinogenicity on a scale from "A" for the greatest evidence to "E" for the least evidence of carcinogenicity in humans (Tables VI, VII, and VIII). The weight-of-the-evidence classification is designed to determine how much available information shows that a compound causes cancer. The evidence need not come from humans, but human risks will be assessed. As described in the narrative of the guidelines, the scientists are encouraged to use their judgment, training, and experience; to consider mechanism of action; to understand and use data on chemical absorption, disposition, and metabolism; to consider the physical and chemical properties of the chemical and the structure-activity relationships for the product; and to weigh other toxicologic effects that may have been at work in the study.

The weight-of-the-evidence criteria, however, were written and sometimes are applied in a fairly prescriptive and mechanical manner. The more studies that show positive results, the greater the weight of the evidence; the greater the weight of the evidence, the higher the qualitative cancer classification; the higher the classification, the more likely a quantitative risk assessment will be conducted, in particular, using the linearized multistage model. Under the current guidelines, EPA scientists are relatively limited in their freedom to adjust the cancer classification on the basis of their judgment, rather than on the number of positive human or (i.e., epidemiological) animal studies. Also, once a quantitative risk assessment has been conducted, the risk manager is likely to rely heavily upon the quantitative risk numbers rather than the qualitative evidence. The risk manager is also inclined to rely on the upper-bound estimate of risk. The guidelines direct scientists to use the qualitative category with an quantitative assessment of risk and to present the uncertainties in the underlying database, but this is rarely done.

The direct relationship between the weight-of-the-evidence categories and the cancer classification is shown in Table VII (<u>12</u>).

Although the chart reproduced in Table VII was included in the cancer guidelines as "Illustrative," the classification system appears to have become more rigid than may have been expected. Agency personnel have often been charged with applying the weight of the evidence in a fairly automatic and mechanical manner. Although this may lead to a simpler or more predictable process to follow when classifying the oncogenic potential of a compound, it does not necessarily lead to more consistent or reliable risk management decisions. The risk assessment should not be conducted without considering the risk management options.

TABLE VI. EPA Classification System for Weight of the Evidence
of Human Carcinogenicity

From Animal Studies:

No Evidence   No increased neoplasms in two well-
                               designed animal studies in different
                               species

No data   No data are available

Inadequate
Evidence   Cannot be interpreted as showing
                               presence or absence of a carcinogenic
                               effect due to major qualitative or
                               quantitative limitations

Limited
Evidence   Data suggest a carcinogenic effect
                               but are limited because: (a) studies
                               involve single species, strain, or
                               experiment; (b) inadequate dosage
                               levels, exposure duration, follow-up,
                               poor survival, too few animals, or
                               inadequate reporting; (c) benign
                               tumors only

Sufficient
Evidence   Increase in malignant or combined
                               benign and malignant tumors: (a) in
                               multiple species or strains; (b) in
                               multiple experiments (different
                               routes or dose levels); or (c) to an
                               unusual degree in a single experiment
                               with regard to high incidence,
                               unusual site or type of tumor, or an
                               early age at onset

SOURCE: Data are from ref. 12.

TABLE VII. Illustrative Categorization of
Evidence Based on Animal
and Human Data

Human Evidence	Sufficient	Limited	Inadequate	No Date	Evidence
Sufficient	A	A	A	A	A
Limited	B1	B1	B1	B1	B1
Inadequate	B2	C	D	D	D
No data	B2	C	D	D	E
No evidence	B2	C	D	D	E

SOURCE: Data are from ref. 12.

TABLE VIII. Categorization of Overall
Weight of the Evidence for Human Carcinogenicity

Group A Human Carcinogen

Group B1 Probable Human Carcinogen with Human Data

Group B2 Probable Human Carcinogen without Human Data

Group C Possible Human Carcinogen

Group D Not Classifiable as to Human Carcinogenicity

Group E Evidence of Non-carcinogenicity for Humans

SOURCE: Data are from ref. 12.

One potential drawback with a mechanical approach is that the risk managers may design certain regulatory responses around the qualitative category (e.g., A, B, or C) and not consider the full weight of the evidence. This has been the case in the EPA Office of Solid Waste, such that the cancer classification determines the reportable quantity for a hazardous waste spill with no further scientific evaluation.

Classification Scheme

The group A carcinogen is a "Human Carcinogen" that must be supported by sufficient human studies (Table VIII). A B2 carcinogen is described as a "probable human carcinogen" by the classification system. The decision to classify a chemical as a B2 carcinogen, however, is made in the absence of human cancer data and rests entirely on the weight of the animal carcinogenesis data. If there is more than one study showing an "increased incidence of malignant tumors or combined benign and malignant tumors (a) in multiple species or strains; or (b) in multiple experiments (e.g., with different routes of administration or using different dose levels); or (c) to an unusual degree in a single experiment with regard to high incidence, unusual site or type of tumor, or early age of onset," then the Agency concludes there is "sufficient evidence of carcinogenicity" (12). These criteria for sufficient evidence fail to provide an opportunity for much scientific judgment and make no reference to mechanism of action, relevance to man, or quality of studies. Sufficient evidence of carcinogenicity in animals, however, automatically leads the Agency to a group B classification and to the label "Probable Human Carcinogen." A B1 carcinogen requires limited human data, regardless of the animal data available. Therefore, "limited" human data lead EPA to conclude that a chemical is a "probable" human carcinogen, rather than possible.

If there are animal data on a pesticide that suggest a carcinogenic effect, the evidence may be sufficient under the guidelines or it may be "limited" if "(a) the studies involve a single species, strain or experiment and do not meet criteria for sufficient evidence. . .(b) the experiments are restricted by inadequate dosage levels, inadequate duration of exposure to the agent, inadequate period of follow-up, poor survival, too few animals, or inadequate reporting; or (c) an increase in the incidence of benign tumors only" (12). Calling the weight of the evidence as limited leads to a group C - Possible Human Carcinogen classification.

Inadequate evidence leads to a group D - Not Classifiable as to Human Carcinogenicity classification. Either no data, insufficient data, or conflicting studies may lead EPA to conclude that the data cannot be interpreted, and a D classification would be appropriate. If "two adequate animal

studies are both negative in different species or in both
adequate epidemiological and animal studies," the group E -
Evidence of Non-Carcinogenicity for Humans is appropriate (12).
    The ultimate result of the current guidelines is that the
more a product is tested, the greater its qualitative cancer
rating. The weight of the evidence increases with any positive
findings, even if the studies are old, of marginal quality,
exceed the maximum tolerated dose, or are shown by mechanistic
studies to be irrelevant to human cancer risk.
    Although mention is made in the guidelines of adjusting the
classification on the basis of supporting information, in
reality very little guidance was provided on when adjustment is
allowed. Therefore, the (EPA) scientists are reluctant to
diverge from a strict reading of the guidelines. The
Pesticides Program has diverged from a strict reading of the
guidelines in some cases, but generally complies with the
system outlined in the EPA Cancer Risk Assessment Guidelines.
Many more pesticides have been classified as group C
carcinogens than might be according to the most conservative
reading of the guidelines.
    One consequence of the guidelines is that they determine
whether or not a quantitative risk assessment is conducted. A
quantitative risk assessment is conducted on all B2 and most
group C carcinogens, whether the data fit the mathematical
model or not. Data sets are often adjusted, for example, by
dropping the results at the high dose level, to fit the
preferred model. Also, because of the structure of the
classification system, many compounds are classified as group C
or B2 carcinogens even if there are data showing that the
compound is metabolized differently in the test species than in
humans, or the results do not appear to model human cancer
risks. The qualitative ranking and quantitative assessment are
used by the risk manager to establish safe levels for worker
and dietary exposure. The EPA Cancer Risk Assessment is very
likely to undergo revision in the near future, but which areas
will be changed is uncertain. (14)
    The EPA Pesticide Programs has demonstrated a willingness
to diverge from a strict application of the guidelines in a
number of cases. One recent example of this is the decision to
terminate the Special Review of linuron (15). Linuron had
increased benign testicular tumors in male rats and benign
liver tumors in male and female mice. Conducting a
quantitative risk assessment on the data would result in an
upper-bound risk estimation as high as one in a thousand based
on worst-case dietary exposure and mixer/loaders without
protective clothing (7). A more realistic estimate of upper-
bound risk was said to be one in ten thousand because of
differences between the worst case and actual exposures. The
Agency decided, however, that it would not regulate linuron on
the basis of oncogenicity. The Toxicology Branch Peer Review
Group, the EPA Carcinogen Assessment Group, and the FIFRA SAP
all agreed that linuron was a group C carcinogen, but they

determined that the qualitative oncogenic potential of linuron was so low that regulating on the basis of oncogenicity would be inappropriate (15).

Pesticides on Food

EPA establishes raw agricultural commodity (RAC) and food additive tolerances for pesticides on food. Carcinogenic pesticides may be registered and tolerances may be set for the raw agricultural commodity if the levels are safe; however, if the concentration of a pesticide on food increases when it is processed, a food additive tolerance is required. The Delaney clause of FFDCA does not allow a food additive tolerance to be set, even if EPA believes it is safe, if the compound has been shown to induce cancer in man or animal. A narrow interpretation of the Delaney clause has prevented registration of new products in circumstances where the upper-bound risks were estimated to be much less than one in a million.

Interactions of FIFRA and FFDCA

The regulation of carcinogenic pesticides on foods depends upon how EPA interprets and applies FIFRA, FFDCA Sections 408 and 409, and especially the Delaney clause in the FFDCA. Pesticide registrations are granted or denied under the FIFRA risk-benefit standard. However, food use pesticides cannot be registered, unless a tolerance or an exemption from a tolerance for the RAC was also granted under Section 408 of the FFDCA. The standard applied in granting a tolerance under FFDCA Section 408 is health-based; that is, the regulation must be set at a level that is not unsafe. Benefits are also considered to some extent in setting the Section 408 tolerance.
    If the RAC, which has been treated with a pesticide, is processed, and if the concentrations of the pesticide in the final form of the processed food are greater than the concentrations found in the RAC, then a food additive tolerance must be established under Section 409 of the FFDCA. Like the 408 RAC tolerance, a Section 409 food additive tolerance must be set at a level that is not unsafe. But over and above the basic requirement of safety, Section 409 prohibits establishing a food additive tolerance when the compound has been shown to induce cancer in man or animals. This prohibition is the Delaney clause.

The Delaney Clause

The Delaney clause can prevent the establishment of tolerances and the registration of pesticides. To explain the effect of the Delaney clause on registration, one can separate pesticides into two groups: (1) where the pesticide has been shown to induce cancer in man or animal (an "oncogenic pesticide") and

(2) where it has not (a "non-oncogenic" or untested pesticide).
If one also assumes that the pesticides in both groups also
otherwise pass the risk-benefit test for FIFRA and FFDCA
Section 408, then it is possible to consider the influence that
the Delaney clause has on pesticide registration.

First, EPA could apply the Delaney clause in a rigid manner
to all pesticides. No distinction would be made between new
and old pesticides. For any oncogenic pesticide, new
tolerances would not be granted, and old tolerances would have
to be revoked. No consideration would be given to the level of
exposure or risk, to the benefits, or to the other hazards of
that product or competing products. This would be very simple
to administer. It would ensure that EPA complies with the
Delaney clause, but it is not necessarily consistent with the
FIFRA.

Second, on the other end of the spectrum, EPA could
completely ignore the Delaney clause. EPA could register
pesticides and set food tolerances for new and old oncogenic
pesticides by considering oncogenicity as part of the
risk-benefit balancing. The FIFRA standard would be complied
with and the FFDCA mandate that the food additive not be
present at unsafe levels could also be satisfied, but it is not
consistent with the Delaney clause. The fact that a pesticide
was an oncogen would not, by itself, preclude registration for
food uses where a food additive tolerance was required. New
and old products would be treated alike with actions measured
on a consistent risk-benefit scale.

Third, the Agency could leave old registrations and
tolerances on the books, even if the data demonstrated that an
oncogenic pesticide needed a food additive tolerance. New
products could be regulated in a much more rigid manner than
the old products, and no food additive tolerances would be
granted for a new oncogenic pesticide regardless of how safe it
was or how low the risks were. This would create a double
standard for old and new pesticides. It may also leave
products on the market with higher risks, when the product
could be replaced by one that posed lower risks but for its
oncogenicity.

The third scenario generally describes the current EPA
approach to pesticide regulation under FIFRA and FFDCA. New
products that are carcinogenic and need a 409 tolerance are not
registered because EPA will not grant a food additive
tolerance. Aliette is an example of this policy. Older
pesticides that are registered and either need or have 409
tolerances have not been cancelled on the basis of Delaney.

Current EPA Approach to Tolerance Setting

Many pesticides were registered before oncogenicity data were
required. Tolerances were established whether or not the
pesticide was oncogenic, because the data were not available.
Those tolerances have not been revoked and registrations have
not been cancelled simply because new data show that a
pesticide may be oncogenic. New tolerances are not set once
the reregistration process begins for an old product. But the

older registrations are essentially grandfathered by the
policy, unless the Agency believes that the risks may be
unreasonable. If the risks are considered unreasonable, the
Agency will initiate a Special Review. But this action is not
taken under Delaney; it is measured under the FIFRA risk-
benefit standard.

New products, if shown to induce cancer in man or animals
and if they need a 409 food additive tolerance, are kept off
the market. This is true even if they replace older and less
safe pesticides. No degree of demonstrated safety has led EPA
to register oncogenic pesticides at this time. This is true
even where the benefits of a new pesticide substantially
outweigh the risks. The FIFRA risk-benefit part of the
equation is not considered as part of the equation if an
oncogenic food-use pesticide needs a 409 tolerance. Older,
less efficacious, and riskier products remain on the market,
however.

Fosetyl al (Aliette) and FFDCA 409

For example, the pesticide fosetyl al (Aliette) is a fungicide
proposed for use on hops. Many fungicides have been shown to
induce tumors in animals and are considered to present
oncogenic risks to humans. Aliette caused an increase in
bladder tumors in rats, but only at very high dose levels
(40,000 ppm). Aliette was classified as a group C carcinogen,
and a quantitative risk estimate demonstrated upper-bound
worst-case risks were less than one in one hundred million.
EPA felt no significant human health risk was posed from this
use. Even though Aliette would replace other much more clearly
carcinogenic fungicides, however, the presence of Aliette on
the processed food would have required a food additive (409)
tolerance. The Delaney clause in Section 409 prevented
establishing a 409 tolerance, because Aliette had been shown to
induce tumors in animals. This strict drafting and
interpretation of FFDCA Section 409 prevented the use of a
beneficial and less risky product.

EPA's Policy Changes with Respect to Delaney

The current policy and legal options were reviewed by the
National Academy of Sciences at the request of EPA. The report
Regulating Pesticides in Food: The Delaney Paradox (13) was
then published by NAS and gives four specific conclusions and
recommendations to the Agency aimed at implementing a
reasonable policy on pesticide residues in food. The Agency,
while criticizing the risk analyses presented in the report,
has publicly embraced the recommendations of the NAS panel.
The Agency does admit that significant legal impediments exist
to implementing them.

The Agency is currently developing an implementation plan
and will publish it in the Federal Register for public notice
and comment. It is expected to draw praise and fire. The plan
is expected to attempt to implement the four recommendations of
the NAS report. Those are:

1. "Pesticide residues in food, whether marketed in raw
    or processed form or governed by old or new
    tolerances, should be regulated on the basis of
    consistent standards. Current laws and regulations
    governing residues in raw and processed food are
    inconsistent with this goal."
2. "A negligible risk standard for carcinogens in food,
    applied consistently to all pesticides and to all
    forms of food, could dramatically reduce total dietary
    exposure to oncogenic pesticides with modest reduction
    of benefits."
3. "The committee's analysis. . .suggests that about 80
    percent of the oncogenic risk from 28 pesticides that
    constitute the committee's risk estimates is
    associated with the residues of 10 compounds in 15
    foods. Logic argues that EPA should focus its
    energies on reducing risk from the most-consumed
    crops, and compelling reasons support such a
    strategy."
4. "The EPA should develop improved tools and methods to
    more systematically estimate the overall impact of
    prospective regulatory actions on health, the
    environment, and food production. Rapid advances in
    computer technology, as well as the EPA's successful
    efforts to computerize major data sets like the
    Tolerance Assessment System (TAS) make such progress
    readily attainable."

The Agency's implementation plan is expected to describe
ways to implement the recommendations of the NAS. The legal
obstacles, however, have been highlighted in the NAS report in
some cases. For example, the first recommendation points out
current laws are inconsistent with treating residues in raw and
processed foods the same. Furthermore, the use of scientific
evidence to demonstrate that a risk is "negligible" is
inconsistent with a strict interpretation of the Delaney
clause. The implementation of the recommendations will pose
legal risks for EPA, and a consistent standard may not be
attained without congressional action.

Negligible Risk

With respect to recommendation 2, EPA apparently will proceed
with a plan that incorporates negligible risk as the target for
setting pesticide tolerances. Negligible risk must be defined
by EPA as part of the implementation plan. The NAS referred to
negligible risk in terms of less than one in a million risk of
cancer. The FDA used a similar range but avoided putting too
precise a point on the issue. EPA is considering several
definitions for de minimis or negligible risk. One approach is
to consider the qualitative nature of the cancer data. For
example, certain mechanisms of action in animals are not good
models for human cancer and may, by definition, be termed
negligible risk even if the quantitative risk assessment showed
risk numbers greater than one in a million (such as group C

carcinogens without a risk assessment or group D carcinogens).
The quantitative risk estimation would be factored into the
determination of negligible risk if the mechanism had not been
shown to be irrelevant and the animal data were sufficient for
a quantitative risk assessment (group B2 or C carcinogens that
do require a quantitative risk assessment).

Regulation of Pesticide Classes

The Agency has routinely regulated pesticides on an active
ingredient by active ingredient basis. Although considering
classes of chemicals would take longer, the Agency could be
expected to reregister or review classes of chemicals. EPA has
used a cluster approach for endangered species regulation. In
the case of endangered species, EPA considered broad categories
of products at the same time (e.g., cropland uses, forestry).
The selection of products reviewed under a project reviewing
chemicals affected by the Delaney clause is likely to be much
narrower in scope (such as tomato fungicides). Although EPA
seemed to accept the recommendation for a categories approach,
it strongly disagreed with the NAS report's characterization of
the actual residues and risks for various crops or pesticides.

Use of Improved Exposure Models

The last recommendation constituted NAS's endorsement of the
EPA tolerance assessment system (TAS). The TAS is a computer-
based system that can analyze risks based on various use
patterns for pesticides. The system greatly improves the
precision of the dietary exposure assessment. TAS is already
being used by EPA and provided the tools used by the NAS in its
own risk analyses.
     If the changes are implemented in the Delaney clause then
the door could be opened to the registration of a number of new
products. The products favorably affected by changes in the
policy would be those that are positive in animal
carcinogenicity studies, but are either of very low potency, or
caused tumors considered less relevant to human risk. The
Delaney clause would not increase the chances of registering a
product that appeared to pose a significant risk by dietary
exposure.
     Changes in the policy would probably also result in the
revocation of tolerances for products on the market with higher
risk numbers. The Agency is likely to eliminate some older,
riskier products, as it registers new products. The
environmental groups will likely also wait for a good test case
and sue the Agency over the setting of a tolerance for a
carcinogenic pesticide.

Other Issues and Conclusion

EPA has regulated many pesticides on the basis of carcinogenic
potential. Dietary residues command a great deal of attention,
but other issues, such as groundwater contamination, consumer
or worker exposure, and Agency resources, also profoundly

affect the regulation of carcinogenic pesticides. Worker exposure is often much greater than dietary exposure and public and Agency concern continues to grow about the presence of even low level carcinogenic pesticides in drinking water. As reregistration generates new data on carcinogenicity, it will remain the focus for future regulations as well. New and more sensitive analytical and toxicological data will identify more compounds to which we are exposed. How EPA assesses and manages cancer risk in the future will profoundly affect producers, users, and the public. Changes will be required in Agency policy and in the pesticide laws to accommodate the regulation of pesticides.

LITERATURE CITED

1. 2,4-D, 2,4-DB, and 2,4-DP; Proposed Decision Not To Initiate a Special Review. Environmental Protection Agency.Federal Register 53:9590-9594, 1988.
2. Decision and Emergency Order Suspending the Registrations of All Pesticide Products Containing Dinoseb. Environmental Protection Agency. Federal Register 51:36634-36650, 1986.
3. Dibromochloropropane (DBCP); Suspension Order and Notice of Intent to Cancel. Environmental Protection Agency. Federal Register 42:57252, 1977.
4. Rebuttable Presumption Against Registration and Continued Registration of Pesticide Products Containing Dibromo-chioropropane (DBCP). Environmental Protection Agency. Federal Register 42:48026-48045, 1977.
5. Suspension Order (DBCP). Environmental Protection Agency. Federal Register 42:57252, 1977.
6. Implementation of Suspension and Conditional Suspension of Registrations of Pesticide Products Containing Dibromochloropropane. Environmental Protection Agency. Federal Register 42:57544-57548, 1977.
7. Emergency Suspension of Ethylene Dibromide (EDB). Environmental Protection Agency. Federal Register 49:4452, 1984.
8. Pesticide Products Containing Heptachlor or Chlordane: Environmental Protection Agency. Federal Register 39:41298-41300, 1974.
9. Consolidated Heptachlor/Chlordane Hearing. Environmental Protection Agency. Federal Register 34:7552-7585, 1976.
10. Consolidated Heptachlor/Chlordane Cancellation Proceedings. Environmental Protection Agency.Federal Register 43:12372-13375, 1978.
11. A Position Document of the U.S. Environmental Protection Agency Office of Pesticide Programs: Selection of a Maximum Tolerated Dose (MTD) in Oncogenicity Studies. U.S. Environmental Protection Agency, 1987.
12. Guidelines for Carcinogen Risk Assessment. Environmental Protection Agency. Federal Register 51:33992-34003, 1986.
13. Risk assessment in the Federal government: Managing the Process. National Research Council. Washington DC, National Academy Press, 1983.

14. Intent to Review Guidelines for Carcinogen Risk
    Assessment. Environmental Protection Agency.   Federal
    Register 53:32656-32658, 1988.
15. Linuron; Preliminary Determination To Conclude the Special
    Review. Environmental Protection Agency.   Federal
    Register 53:31263-31268, 1988.
16. Regulating Pesticides in Food: The Delaney Paradox.
    National Research Council.   272pp, 1987.
17. Office of Pesticide Programs.   Fiscal Year 87 Report on
    the Status of Chemicals in the Special Review Program,
    Registration Standard Program, Data Call-In Programs, and
    Other Registration Activities.   Environmental Protection
    Agency.   NTIS PB88-158886, Springfield, VA.

RECEIVED August 8, 1989

Chapter 3

Pathogenesis of Neoplasia and Influences of Pesticides

Gary M. Williams

American Health Foundation, 1 Dana Road, Valhalla, NY 10595

The evolution of a neoplasm is a complex multi-event and multi-stage process which proceedes through two sequences, the conversion of normal cells to neoplastic cells and the development of neoplastic cells into tumors. These sequences have been documented experimentally to be similar in a number of tissues affected by chemical carcinogens. The cellular events in experimental carcinogenesis also have their counterparts in human cancer development. Chemicals, including pesticides, affect the carcinogenic process in a variety of ways, both facilitory and inhibitory, in the sequences of neoplastic conversion and development. Some chemicals exert more than one efffect on the neoplastic process.

The pathogenesis of chemically-induced cancer is complex, consisting of a series of events. The process can be divided into two distinct sequences, neoplastic conversion, involving change in the genetic apparatus of cells leading to generation of a neoplastic cell and neoplastic development in which the neoplastic cell evolves into a tumor (Fig 1). In experimental models, pesticides can influence both sequences, either enhancing or inhibiting them.

Chemical Carcinogens

Chemical carcinogens are defined operationally by their ability to produce an increase in tumor incidence. Chemicals capable of eliciting a tumor response in experimental animals comprise a highly diverse collection of structural types of chemicals (1). Considering this fact alone, it has seemed likely that the tumorigenic effects of carcinogens could be exerted by several mechanisms. Evidence of this is provided by observations that, al-

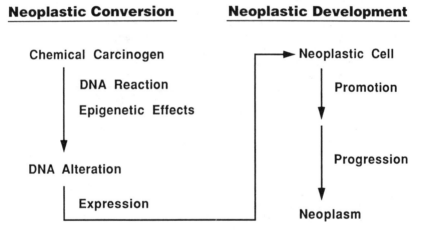

Figure 1.   Outline of the carcinogenic process.

though many carcinogens give rise to reactive species that damage DNA, some carcinogenic chemicals, notably hormones, do not have these properties.

In recognition of this fundamental difference, a mechanistic categorization of carcinogens into two main types, DNA-reactive and epigenetic, (Table I) has been developed (2).

Table I.   Classification of Carcinogens

Type of Carcinogen	Example of type
DNA-reactive (Genotoxic)	
Activation-independent	Alkylating agent
Activation-dependent	Polycyclic aromatic hydrocarbon, nitrosamine
Inorganic	metal
Epigenetic	
Promoter	organochlorine compounds saccharin, drugs
Hormone-modifying	amitrole
Immunosuppressor	purine analog
Cytotoxic	nitritotriacetate
Peroxisome proliferator	lactofen
Unclassified	methapyrilene

The distinction between DNA-reactive or genotoxic carcinogens and the epigenetic type is important to the understanding of the influences of pesticides on the pathogenesis of neoplasia, since there are major differences between the two types in chemical actions and mechanisms.

The Neoplastic Process

The neoplastic phenotype is transmitted to the progeny of neoplastic cells and thus must involve a change in the structure or expression of genetic information. DNA-reactive carcinogens are capable of effecting such an alteration directly through a mutational event, either in base sequences or gene arrangement. In contrast, epigenetic agents may act either by facilitating expression of a preexisting abnormal genome or by inducing an abnormal genome through: 1) spontaneous mutation during increased levels of induced cell proliferation; 2) induced mutation through impairment of the fidelity of DNA polymerases; 3) induction of a stable altered state of gene expression; or 4) generation of intracellular reactive species such as activated oxygen, which are in turn genotoxic.

The sequence of events by which chemicals produce cancer from their action on normal cells is outlined in Figure 1.

Biotransformation

    Certain synthetic DNA-reactive carcinogens have chemical
reactivity inherent in their structures and are referred to as
direct-acting, primary, or activation-independent carcinogens.
Apart from these, most DNA-reactive carcinogens require
biotransformation by host enzyme systems into reactive metabolites
and, accordingly, are designated as indirect, secondary, procarcinogens or activation-dependent carcinogens. Biotransformation
results from the operation of enzyme systems involved in the metabolism of endogenous substrates, but which can also act on
xenobiotics. The principal enzymes that biotransform chemicals are
part of the cytochrome P-450 dependent monooxygenase system associated with the endoplasmic reticulum. For all classes of
genotoxic carcinogens, except nitrosamines, most metabolic steps
generate detoxified water-soluble metabolites which can be excreted, usualy in the form of conjugates. Thus, the metabolism of
carcinogens to activated forms is a minor byproduct of biotransformation.
    One reaction, sulfoconjugation, is detoxifying for C-OH com-
pounds, but activating for N-OH compounds. A few activation reactions, such as N-oxidation, acetylation, or nitro reduction can be
performed by enzymes other than those of the cytochrome system.
Most biotransformation takes place in the liver, with other organs
involved to varying degrees (3).
    Certain bacterial enzymatic actions, such as glucoside
cleavge, nitro reduction, and azo reduction for certain tetrazo
dyes, are involved in the activation of compounds in the intestine.
    For many epignetic carcinogens, such as hormones and or-
ganochlorine pesticides, metabolism leads to detoxified products.
However, an activation reaction metabolizes certain immunosuppressants to their cytotoxic forms. It has been suggested that induction of a P-452 oxidation of fatty acids may underlie the ability
of agents to induce peroxisomes (4). The latter event, is believed
to underlie the carcinogenicity of peroxisome proliferators (5),
among which are some pesticides (6).
    Major differences exist between species in biotransformation
processes. For example, most animal species display either rapid
(e.g. hamster) or slow (e.g. rat) acetylation activity, whereas
humans are endowed with a genetically determined polymorphism (7).
Differences in biotransformation activities account for many differences in susceptibility to carcinogenesis. For example, it has
been demonstrated that differences in acetylation activity influence the genotoxicity of aromatic amines (8).
    A variety of chemicals modify biotransformation processes
(9). Among these are a number of organochlorinc pesticides, such
as DDT, which induce liver enzyme activites. In general, such pesticides inhibit the carcinogenicity to experimental animals of
activation-dependent carcinogens.

Ultimate Carcinogen

For organic direct-acting carcinogens and activated metabolites of procarcinogens, the ultimate reactive form of these genotoxic carcinogens is an electrophilic reactant (10). For epigenetic carcinogens, it is also possible that reactive species could be generated from normal cellular consitituents. Hydrazine has been reported to give rise a methylating species (11). Peroxisome proliferators appear to lead to generation of reactive oxygen species as a result of production of H_2O_2 during oxidation of lipids in peroxisomes (5).

Macromolecular Interactions

The ultimate electrophilic forms of DNA-reactive carcinogens can react covalently with nucleophilic sites in protein, RNA and DNA (12). Glutathione is also a good nucleophile in competition with other molecules. Binding to proteins, because of their relative abundance in cells, is usually the major macromolecular interaction of carcinogens. As a result of reaction with DNA, carcinogens of this type are mutagenic and active in other short-term tests for carcinogens (13).

The ultimate electrophilic reactants of carcinogens can bind to all four bases of DNA as well as to the phosphodiester backbone (14). The base adducts are formed at several sites, with the most susceptible site appearing to be the purine nitrogen, e.g., nitrosamines alkylate guanine at the N7 position and, to a lesser extent, the O6 position. Aflatoxin B_1 binds at the N7 position of guanine, 2-acetylaminofluorene interacts at the C8 and N2 positions, and benzo(a)pyrene is bound to the N2 position. Biomonitoring approaches have demonstrated DNA-bound products in humans exposed to enviornmental genotoxic carcinogens (13).

Considerable evidence now indicates that binding to DNA is a critical reaction of carcinogens (1,14) as implied in the classification of such carcinogens as DNA-reactive or genotoxic. With alkylating agents, O6 alkylation appears to be highly relevant to the carcinogenic effect, and with benzo(a)pyrene the binding of the trans-7,8-dihydrodiol-9,10-epoxide to guanine seems to be the key reaction. The particular regions of DNA i.e., the specific genes whose modification is essential to initiation of the carcinogenic process are begining to be identified, as discussed below.

Several carcinogenic pesticides are known to be DNA-reactive, for example, ethylene dibromide. Most, however lack this activity (15).

For some types of epigenetic carcinogens, such as hormones (Table 1), noncovalent binding to specific cellular receptors is undoubtedly essential to their oncogenic effects. Effects on cell membranes may underlie the action of carcinogens that operate as tumor promoters. For the organochlorine pesticides that act as tumor promoters (see below), such effects on cell membranes may be critical.

DNA Repair

The damage produced in DNA by genotoxic carcinogens can be
corrected, primarily by repair process in which the damage or the
region containing the damage is removed (14).

In the excision type of repair, an incision is made by an en-
donuclease in the DNA strand in the vinicity of the DNA damage, a
stretch of DNA containing the region of damage is excised, and a
patch is synthesized, followed by rejoining of the strand. One
type of excision repair, nucleotide excision is elicited by the introduction of bulky adducts into DNA and may result in the removal
of 80-100 nucleotides per adduct. Another type, base excision is
provoked by smaller modification of bases such as alkylation and
usually involves the removal of only 3-4 bases surrounding the
damaged region. An additional type of repair is the removal the
alkyl group from O6-alkylguanine by the O6-alkylguanine-DNA-
alkyltransferase system.

Different types of damage are repaired at different rates;
thus, the C8-guanine adduct of 2-acetylaminoluorene and other
aromatic amines is removed at a considerably faster rate than the
N2 adduct, perhaps because the latter does not lead to major
denaturation of the double helix. In addition, significant species
and tissue differences in rates of repair exist. For example,
damage to liver DNA by dimethylnitrosamine is more slowly repaired
in the hamster than in the rat. In mice and rats, the residues of
O6-alkylguanine produced by ethylnitrosurea in liver and kidney are
rapidly removed, whereas alkylation is highly persistent in brain.
The activity of the alkyltransferase system is inducible.

In general, humans are more proficient in DNA repair
processes than animals, although genetically-determined deficiency
states, such as xeroderma pigmentosum, occur.

Altered Effector

Considerable evidence now indicates that the effector for
neoplastic conversion of cells is DNA. If the damage to DNA by a
genotoxic carcinogen is not repaired and the affected region is
used as the template for synthesis of new DNA, a permanent mutation
can be introduced through mispairing of bases. Because of this
rapidly proliferating tissues and those stimulated to proliferate
are highly susceptible to carcinogens. Several types of damage to
DNA are now known to be promutagenic (16). For example alkylation
of the O6 position of guanine results in base pairing with thymine
rather than cytosine (17).

The nature of the permanently altered effector that is critical to neoplastic conversion increasingly appears to involve activation of cellular oncogenes (14,18,19), see below. Specific
mutations in oncogenes have been identified (19), as well as other
changes. These occur in oncogenes isolated from neoplasms arising
in both experimental animals and humans (20,21).

Co-Carcinogenicity

The phenonmenon of co-carcinogenicity is the enhancement of carcinogenicity of a chemical by another concurrently administered
chemical, which, under the test conditions, is not itself car-

cinogenic (22). Several mechanisms are possible for the action of co-carcinogens and it seems likely that different mechanisms operate in specific situations. A pesticide operating as a co-carcinogen has not been described.

Agents believed to operate as co-carcinogens in humans include tobacco smoke and alcohol (23). No evidence exists for a co-carcinogenic action of pesticides in humans.

Neoplastic Cells

Neoplastic cells have an altered genome, including both DNA and chromosomal mutation. The essential biological abnormality in neoplastic cells is a loss of growth control. Certain lines of evidence suggest that the neoplastic state is a homozygous recessive condition (24), which may reflect inactivation of anti-oncogenes. Other studies point to the activation of dominant cellular oncogenes (25). The function of anti-oncogenes is not understood, but it is well-established that oncogenes code for factors involved in cellular growth (26,27).

The neoplastic cell may persist in a dormant state for months, in spite of genetic alterations, or, depending upon host conditions, grow to form a neoplasm. The elements which prevent expression of initiated cells as neoplasms are not understood, but may involve factors, such as chalones, which regulate growth and differentiation. Large molecules can be exchanged between cells through specialized membrane structures known as gap junctions. Thus, transmission of regulatory factors from normal to initiated cells may effect control of the latter. The cells of fully developed neoplasms are deficient in gap junctions and possess other membrane abnormalities, indicating a limitation in their ability to receive regulatory signals.

Promotion

The classical definition of promotion is the enhancement of the carcinogenicity of an agent by a second agent, not carcinogenic by itself under the test conditions, acting after exposure to the first has ended (22). In experimental animals, promotion has been shown to occur in most organs, including skin, liver, stomach, colon, breast, and bladder (28).

Although promoters are usually regarded as being noncarcinogenic, most will in fact elicit tumor formation albeit in small yield, when administered alone under conditions of prolonged exposure at high levels. This is probably the basis for the carcinogenicty of agents such as saccharin and certain organochlorine pesticides (Table I).

A variety of effects have been suggested to underlie the promoting action of chemicals (9). Since initiated cells can remain dormant in tissues for many months, it seems evident that these altered cells are being kept under some kind of growth regulation. As described above, cells exchange lare molecules through membrane gap junctions. If this kind of intercellular exchange is involved in the regulation of differentiation and growth, then interference with this process could release dormant tumor

cells for growth into neoplasms. Thus, tumor promotion may be explained by interference with the growth control suppression of
latent tumor cells. A large number of tumor promoters have now
been demonstrated to have the ability to inhibit intercellular communication (29), and therefore, this effect is assuming importance
as one basis for tumor promotion.

Several organochlorine pesticides have been documented to be
liver tumor promoters in experimental animals (Table II). An interesting species difference is that hamsters were found to be
resistant to DDT promotion (32).
Agents believed to operate as promoters in humans include
tobacco smoke, hormones and bile acids (23). No evidence exists
for a promoting action of pesticides in humans.

Table II. Liver Neoplasm Promotion by Organochlorine
Pesticides

Pesticide	Species	Effect	Reference
DDT	rat	+	30
	mouse	+	31
	hamster	-	32
Chlordane	mouse	+	31
hepatchlor	mouse	+	31

Progression

Neoplasms can undergo permanent stable changes in their
phenotype, a process referred to as progression (33). Little is
known about the basis for this alteration in the characteristics of
neoplasms. It could result from gene amplification or a change in
their chromosomal complement. Another hypothesis (34) is that
decreased fidelity of DNA polymerases in tumor cells leads to errors in the replication of DNA, thereby introducing new mutations.

Conclusions

As detailed, the overall carcinogenic process is complex and
involves a series of steps comparising two distinct sequences.
Basically, the process is similar in experimental animals and in
humans. In fact, several chemical carcinogens have been shown to
exert similar effects, such as the type of DNA adduct and type of
neoplasm, in both animals and humans. Nevertheless, there are
definite quantitative differences between species of animals and
between experimental animals and humans. These differences make
simplistic mathematical extrapolation of animal data to potential
human effects a non-scientific enterprise (35).
Most chemicals that have caused cancer in humans are of the
DNA-reactive type (35-37). Given sufficient exposure, it seems
likely that any experimental carcinogen of this type would produce
cancer in humans. Many have not (38), however, which may be due to

the defense mechanisms (i.e. chemical detoxification and DNA repair processes) with which humans are endowed. Regardless, carcinogens of this type should be regarded as qualitative hazards (36).

Few experimental carcinogens of the epigenetic type have been associated with cancer in humans, these are mainly hormones or immunosuppressants. Humans have been exposed to many pesticides known to cause cancer in experimental animals, but none has been linked to cancer in humans (38). The absence of effects in humans has been suggested to be due the fact that exposures of humans are below the threshold for the biological effect (e.g. peroxisome proliferation or membrane alteration and consequent promoting action) underlying carcinogenicity (36). Moreover, some of the conditions necessary for tumor induction in rodents may not be attainable or tolerable to humans.

References

1. Searle, C.E. Chemical Carcinogens; American Chemical Society:Washington, DC, 1984; 2nd ed., ACS Monograph 182.
2. Williams, G.M.; Weisburger, J.H., In Toxicology. The Basic Sciences Of Poison; Klaassen, C., Amdur, M., Doull, J.; MacMillan:New York, 1986; 3rd ed. p. 99.
3. Weisburger, J.H.; Williams, G.M., In Cancer: A Comprehensive Treatise; Becker, F.F. Ed.; Plenum Press:New York, 1982; 2nd ed., p. 241.
4. Sharma, R.; Lake, B.G.; Foster, J.; Gibson, G.G. Biochem. Pharmacol. 1988, 37, 1193-1201.
5. Rao, M.S.; Reddy, J.K. Carcinogenesis 1987, 8, 631-36.
6. Butler, E.G.; Tanaka, T.; Ichida, T.; Maruyama, H.; Leber, A.P.; Williams, G.M. Toxicol. Appl. Pharmacol. 1988, 93, 72-80.
7. Weber, W.W. The Acetylator Genes And Drug Response; Oxford University Press:New York, 1987.
8. McQueen, C.A.; Maslansky, C.J.; Williams, G.M. Cancer Research 1983, 43, 3120-23.
9. Williams, G.M. Fundamental and Applied Toxicology 1984, 4, 325-44.
10. Miller, E.; Miller J. In Origins of Human Cancer; Hiatt, , Watson, , Winsten, , Eds.; Cold Spring Harbor Laboratories:New York, 1977; p. 605.
11. Bosan, W.S.; Shank, R.C. Toxicol. Appl. Pharmacol. 1983, 70, 324-34.
12. Williams, G.M.; Weisburger, J.H. In A Guide to General Toxicology; Homburger, F., Hayes, J.A. Eds.; Karger:New York, 1983; Chapter 12.
13. Williams, G.M. Ann. Rev. Pharmacol. Toxicol. 1989, 29, 189-211.
14. Weinstein, I.B.; Vogel, H.J. Genes and Proteins in Oncogenesis; Academic Press:New York, 1983.
15. Williams, G.M. In The Pesticide Chemist and Modern Toxicology. Bandal, S.K., Marco, G.J., Golberg, L. and Leng, M.L. Eds.; American Chemical Society Symposium Series 160; American Chemical Society:Washington, DC, 1981; p. 45.

16. Rossman, T.G.; Klein, C.B. Environ. Molec. Mutagen. 1988, 11, 119-33.
17. Snow, E.T.; Foote, R.S.; Mitra, S. J. Biol. Chem. 1984, 259, 8095-100.
18. Alitalo, K.; Koskinen, P.; Makela, T.P.; Saksela, K.; Sistonen, L.; Winquist, R. Biochimica et Biophysica Acta 1987 907, 1-32.
19. Bos, J.L. Mutation. Research. 1988,, 195, 255-71.
20. Der, C.J. Clin. Chem. 1987, 33, 641-46.
21. Nishimura, S.; Sekiya, T. J. Biochem. 1987, 243, 313-27.
22. Berenblum, I. Carcinogenesis as A Biological Problem; Amreican Elsevier:New York, 1974.
23. Williams, G.M. In Cancer of the Respiratory tract: Predisposing Factors; Mass, M.J., Kaufman, D.G., Siegfried, J.M., Steele, V.E., Nesnow, S., Eds.; Raven Press:New York, 1985; p. 447
24. Harris, H. Cancer Research 1988, 48, 3302-6.
25. Weinberg, R.A. Cancer 1988, 61, 1963-8.
26. Heldin, C.H.; Betsholtz, C.; Claesson-Welsh, L.; Westermark, B. Biochimica et Biophysica Acta 1987, 907, 219-44.
27. Deuel, T.F. Ann. Rev. Cell. Biol. 1987, 3, 443-92.
28. Slaga, T.J.; Sivak, A.; Boutwell, R.K. Carcinogenesis: A Comprehensive Survey. Mechanisms of Tumor Promotion and Carcinogenesis; Raven Press:New York, 1983.
29. Barrett, J.C.; Kakunaga, T.; Kuroki, T.; Neubert, D.; Trosko, J.E.; Vasiliev, J.M.; Williams, G.M.; Yamasaki, H. In Long-Term and Short-Term Assays for Carcinogens: A Critical Appriaisal Montesano, R., Bartsch, H., Vainio, H., Wilbourn, J., Venitt, S. Eds.; IARC:Lyon, France, 1986, p. 287.
30. Peraino, C.; Fry, R.J.; Staffeldt, E.; Christopher, J. P. Cancer Res. 1975, 35, 2884-90.
31. Williams, G.M.; Numoto, S. Carcinogenesis 1984, 5, 1689-96.
32. Tanaka, T.; Mori, H.; Williams, G.M. Carcinogenesis 1987, 8, 1171-78.
33. Foulds, L. Neoplastic Development; Academic Press:New York, 1969; Vol. 1.
34. Springgate, C.F.; Leob, L.A. Proc. Nat. Acd. Sci. USA 1973, 70, 245-9.
35. Williams, G.M.; Reiss, B.; Weisburger, J.H. In Mechanisms and Toxicity of Chemical Carcinogens and Mutagens; Flamm, G, Lorentzen, R. Eds.; Princeton Scietific Publishers:Princeton, NJ, 1985; Vol. 11, p. 207.
36. Williams, G.M. In Banbury Report 25: Nongenotoxic Mechanisms in Carcinogenesis. Cold Spring Laboratory, 1987; p. 367. Williams, G.M.; Numoto, S. Carcinogenesis 1984, 5, 1689-96.
37. Bartsch, H.; Malaveille, C. Cell Biol. Toxicol. 1989, 5, 000-000.
38. International Agency for Research on Cancer. IARC Monographs on the Evaluation of Carcinogenic Risks to Humans Overall Evaluations of Carcinogenicity: An Updating of IARC Monographs, Volumes 1-42, IARC:Lyon France, 1987.

RECEIVED August 14, 1989

Chapter 4

Mechanisms of Chemical Carcinogenicity

W. T. Stott, T. R. Fox, R. H. Reitz, and P. G. Watanabe

Mammalian and Environmental Toxicology Research Laboratory, Health and Environmental Sciences, 1803 Building, The Dow Chemical Company, Midland, MI 48674

Chemicals may cause tumors in animals via a
number of very distinct mechanisms of action.
Based upon the ability to interact with
genetic material, a general classification
scheme of chemical carcinogens has evolved.
In general, these compounds fall into two
broad categories: 1) genotoxic chemicals,
which have potential to directly damage
genetic material and cause mutations; and 2)
nongenotoxic chemicals, whose primary
mechanism(s) of action involve an interrelated
sequence of identifiable physiological and
biochemical changes. The practical and
theoretical basis for this classification
scheme, how several pesticides exhibiting
different mechanisms of action fit into it and
the significant impact that this information
may have in terms of risk assessment is
presented.

Since the earliest observations of chemical
carcinogenesis in animals a rather obvious question has
confronted cancer biologists. What is the potential
carcinogenic risk to humans from compounds, be they of
"natural" or synthetic origin, which have been shown to
cause cancer in other animal species? The answer to this
question is of course dependent upon the
interrelationship of exposure and toxicity. One without
the other does not represent risk, in this case a
carcinogenic risk to humans. This relatively simple
principle is as true at the cellular/molecular level as
it is at the macroscopic, whole animal level where it has
particular significance for the application of

0097–6156/89/0414–0043$09.75/0
© 1989 American Chemical Society

agrochemicals. However, assuming equal exposure, do all
chemicals which have been shown to be tumorigenic in
animal bioassays pose the same carcinogenic risk? The
answer to this question lies to a major degree in the
understanding of the mechanisms which drive the
neoplastic process itself at the most fundamental level.
This knowledge, and the models evolving to exploit it,
may have practical application in the risk assessment
process. When coupled with appropriate metabolism and
pharmacokinetic data, a more comprehensive, wholistic
conceptualization of chemical carcinogenesis may be
achieved and utilized in the risk assessment of
tumorigenic chemicals.

This paper explores the theoretical and experimental
basis for both the practical application of mechanistic
data to the interpretation of rodent bioassay results as
well as the development of a biologically-based, more
wholistic, model of chemical carcinogenesis.

STAGES OF NEOPLASIA

Tumorigenesis is generally recognized as a complex,
multi-step process involving a multitude of interrelated
changes in gene expression, cellular physiology and
biochemistry. However, tumor biologists typically define
tumor development as occurring in three discrete stages
based upon morphological and molecular characteristics;
initiation, promotion and progression. Briefly, the
initiation stage defines the event(s) which begin the
transformation of a single progenitor cell to a
multicellular malignant tumor. Initiation is believed,
in most instances, to involve a permanent, irreversible
change probably consisting of a genetic mutation. The
initiated cell now has the potential to develop into a
tumor providing that the other essential steps of
carcinogenesis take place, promotion and progression. If
these events do not occur the intiated cell may remain
essentially dormant indefinitely with no apparent adverse
effects to the host. Promotion represents a period during
which, given the proper conditions, there is a clonal
expansion of the initiated cell(s) within a given tissue.
Promotion can be thought of as increasing the population
of target cells which are then available for the further
changes to take place which are necessary for the
complete malignant conversion of the initiated cell.
These changes demarcate progression, which can be viewed
as a tumor evolutionary event. Additional genetic (point
mutations, chromosomal rearrangements, additions and
deletions) and nongenetic changes will occur until the
growing tumor consists of a heterogeneous population of
cells. Ultimately, one or more cells may emerge from
this process having acquired all the necessary attributes
of a completely neoplastic cell.

The multi-stage cancer model serves as a paradigm to recognize the functional importance of the major critical events of tumorigenesis. Although promotion and progression have been defined as distinct and separate events of carcinogenesis, they may not necessarily occur in a sequential stepwise manner. It is entirely feasible that biochemical and genetic changes associated with progression may also be occurring in some cells during the rapid cellular proliferation of the promotion phase.

MECHANISM OF CARCINOGENESIS

Somatic Mutation Theory. The initial event (initiation) in the process of neoplastic transformation of a cell which may ultimately lead to tumor formation is widely held to involve a mutation in the DNA of a critical gene of the genome of that cell. This concept, originating as the "somatic mutation theory of carcinogenesis" as outlined by Boveri in 1929 (1), forms the logical basis for our present day understanding of the mechanism of chemical carcinogenesis. Simply stated, this theory dictates that mutations in critical sites of the genome of a somatic cell (i.e. DNA) as a result of chemical interaction may, in turn, result in the anaplastic transformation of the cell. The classic example of this "interaction" is a fairly straight forward chemical reaction involving the attack and alkylation of DNA bases by an electrophilic molecule or the intercalation and hydrogen or covalent binding of a more planer molecule within the helical matrix of the DNA. Several examples of reaction products of DNA bases with halo-(short-chained)nitrosoureas are shown in Figure 1. The resultant changes in the bases or physical disruption of the helical DNA structure or both can result in mismatched base pairing during replicative synthesis. Several examples of the mutations which can occur as a result of DNA alkylation and their altered phenotypic expression (i.e. altered amino acid incorporation into the protein product) are shown in Table 1.

A mitigating factor in this process are the various cellular DNA repair enzyme systems which may remove the chemically induced lesion before or after DNA replication. A number of these which actively repair damage as a result of short-chained alkylation, much as would be expected to occur as a result of exposure to the genotoxic pesticides methyl bromide or ethylene dibromide, are listed in Table 2. However, the activity of repair enzyme systems does represent a saturable process and in some instances may itself be the source of errors in DNA base sequences, possibly by an inducible error-prone DNA repair enzyme system analogous to the so-

Table 1. Examples of chemically induced mutations
(adapted from Topal (2))

Compound	Base Change[a]	Amino Acid
N-hydroxy-2-acetylaminofluorene		
	C to A	glutamine to lysine
7,8-dimethylbenz(a)anthracene		
	A to T	glutamine to leucine
vinyl chloride	A to T	glutamine to leucine
	A to G	glutamine to arginine
1'-hydroxy-2',3'-dehydroestragole		
	A to G	glutamine to arginine

[a]Definitions: C, cytosine; T, thymidine; A, adenine; G,
guanine

called SOS repair observed in bacteria. Another factor
is the presence of innumerable noncritical sites within a
cell such as proteins, sugars, fatty acids and even DNA
itself which offer alternate targets for a reactive
molecule. Covalent binding of reactive chemicals with
these abundant molecules may, at best, result in little
or no change in the function of the alkylated molecule
and, at worst, result in cell death. These reactions in
toto, however, serve to effectively dilute the critical
target sites within a cell.

Table 2. Examples of repair mechanisms for alkylated DNA
bases and their substrate specificity (adapted from
Ludlum and Papirmeister (3))

Repair Process	Substrate
Excision Repair	
- uvrabc excision	Crosslinks or helical distortions nuclease*
- DNA glycosylases	N7-substituted guanines (with or without ring opening)
N3-alkyladenines	
- AP endonucleases	Depurinated sites
Transferase Repair	O6-alkylguanines
RecA-dependent Repair*	Phosphotriesters

*prokaryote

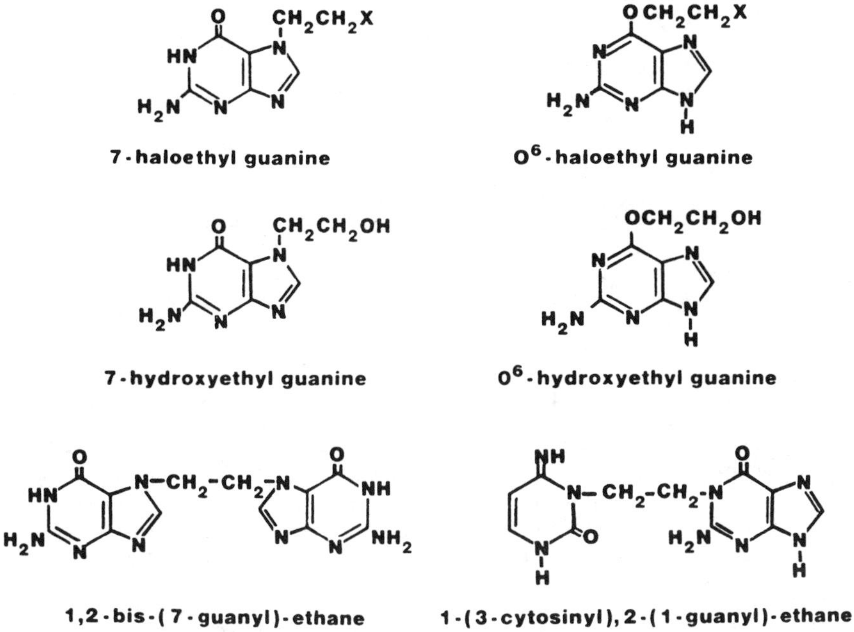

7-haloethyl guanine

O^6-haloethyl guanine

7-hydroxyethyl guanine

O^6-hydroxyethyl guanine

1,2-bis-(7-guanyl)-ethane

1-(3-cytosinyl),2-(1-guanyl)-ethane

Figure 1. Examples of DNA base modifications which may occur as a result of base alkylations by halo-(short-chained)-alkylnitrosamides.

Figure 2 displays the possible kinetics of DNA alkylation and repair based upon a model presented by Gehring and Blau (4). In this model, all rate processes are considered to be saturable and dose-dependent; activation of the compound, elimination, covalent binding to noncritical and critical sites, and repair of critical site alkylation. If each of the respective parameters were not dose dependent and thus not saturable, the levels of noncritical binding, repaired molecules, and critical binding to genetic material escaping prereplicative repair would be parallel to the X-axis. Instead, it can be seen that at a relatively high, saturating dose level, a disproportionate increase in the quantity of covalent binding of the activated chemical to critical molecules occurs as repair becomes saturated. This simulation serves to emphasize the relatively straight forward interrelationships of dose-dependent metabolism and repair mechanisms of a cell in defining a true level of reactive molecule interaction with genetic material. Indeed, it has been suggested that the tumorigenic potency of animal carcinogens may be estimated based upon the quantitative covalent binding of these chemicals to DNA in vivo (5) and that utilizing a derivation of this data, called the Covalent Binding Index, carcinogens may be categorized as to their potential carcinogenic risk. Several examples demonstrating this correlation are given in Table 3.

Alternate Models. As reviewed by Trosko and Chang (6), there are considerable experimental data to support the mutational origin of tumorigenesis as a result of chemical interaction. These data include findings regarding the clonal nature of tumors, the bacterial and eukaryotic cell mutagenicity of many carcinogens, the correlation of high cellular mutation rates and skin cancer in individuals lacking normal DNA repair systems, and the experimentally demonstrated involvement of mutation in the initiation phase of some hydrocarbon-induced carcinogenesis. However, some experimental data do not support a direct mutational origin of carcinogenesis. These data have included the induction of tumors in animals following plastic or metal film implantation or hormonal imbalance (7), the apparent totipotency of some tumor cell genomes (8-10), and the observation that not all carcinogens appear to be mutagenic in standard short-term assays designed to detect such somatic mutations.

Several theories have been proposed which reconcile these different mechanisms of carcinogenesis either by

Table 3. Correlation of hepatocarcinogenicity of
chemicals in the rat with the covalent binding index for
rat-liver DNA (adapted from Lutz (5))

Compound	CBI^a
Strong Hepatocarcinogens	
Aflatoxin B1	17,000
Dimethylnitrosamine	6,000
Diethylntrosamine	42-430
Moderate Hepatocarcinogens	
2-Acetylaminofluorene	560
Vinyl Chloride	525
Nitrosopyrolidine	180
Ethylene Dibromide	180
Weak Hepatocarcinogens	
Urethane	29-90
para-Aminoazobenzene	2
Saccharin	<0.005

^{a}CBI = (#alkylations/10E6 nucleotides) / (mmol/kg dosage)

proposing an alternative to mutagenesis or by
accommodating both mutational and nonmutational
mechanisms. Most notable among these has been the
"integrative theory of carcinogenesis" proposed by Trosko
and Chang (6) which encompassed both mutational and
nonmutational origins of carcinogenesis and a model
proposed by Moolgavkar and Knudson (11) which was a
multi-hit, more mathmatically-oriented, theory for
predicting cancer rates in humans. Both provided useful
paradigms for the study of the molecular events leading
to cancer. In Trosko and Chang's (6) theory it was
proposed that neoplasia may arise; 1) from a mutagenic
event in a regulatory locus alone if associated
structural genes are in a transcribable state, 2) from a
mutagenic event if there is a coupled nonmutational
alteration of the genes into a transcribable state
(promotion), or 3) by the "abnormal" derepression of
genes during critical developmental states which prevent
normal gene regulation (e.g. hormone-induced

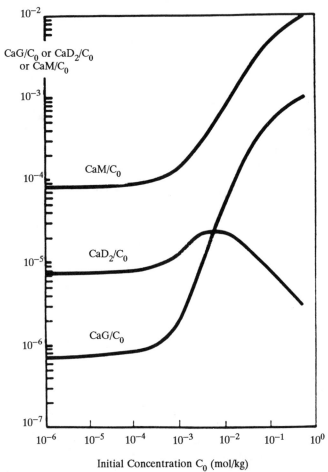

Figure 2.   Simulated kinetics of binding of a
metabolically activated genotoxin to genetic material
(CaG) and total macromolecules (CaM).   The kinetics of
metabolite detoxification by a saturable repair process
(CaD$_2$) is also displayed.   All values have been
normalized to the initial concentration of the compound
(C$_0$).

carcinogenesis). In contrast, Moolgavkar and Knudson's (11) model was based upon epidemiological data for lung and breast cancer in humans and incorporated the transition of target stem cells into cancer cells via an intermediate stage in two irreversible steps. This model appeared to fit the experimentally demonstrable two-stage carcinogenesis model and was consistent with the development of homozygosity at a "cancer gene" locus. Further, it provided a framework for understanding the effects of age upon cancer incidence; the increased susceptibility of certain populations and individuals (i.e. genetic predisposition); and the results of animal carcinogenesis bioassays. In both theories a critical site within the genome was still envisioned to be a necessary target for mutagenic chemicals if they were to cause tumors in animals. An important advance has been the discovery of what appear to be these critical sites within the genome, proto-oncogenes.

ONCOGENES, A UNIFYING FACTOR?

Background. Efforts to understand the molecular mechanisms of cancer were given a tremendous boost by the discovery in the early 1980's of several genes which appeared to represent the long envisioned critical target sites for mutagenic activity within the genome of mammalian cells (12). Termed cellular "proto-oncogenes", these genes appear to play an essential role in normal cellular proliferation and differentiation and have been highly conserved throughout evolution. However, qualitative changes in these genes, for example via point mutations, or quantitative changes, for example via chromosomal rearrangements resulting in over expression, can facilitate the transformation of a normal cell into a cancerous one (13, 14). An example of this activation process is the transformation of the H-ras proto-oncogene to its oncogenic form. This activation involves point mutations within specific regions of the gene. For example, in human bladder cancer cells this mutation may occur in codon 12 which results in the amino acid substitution of glycine for valine in the gene product. In contrast, the activation of c-myc proto-oncogenes in Burkett's lymphoma patients involves a chromosomal translocation event. The c-myc gene normally resides on chromosome 8 but is found in its activated state upon chromosome 14. The rearrangement of this gene leads to abnormal expression which plays an important etiological role in this type of cancer. Additional work upon the activation of cellular proto-oncogenes by chemicals will continue to elucidate the specific role that these agents play in the induction of cancer.

In vitro DNA transfection studies using NIH 3T3 cells
have detected activated cellular proto-oncogenes in a
variety of human tumors and tumor cell lines (15). Figure
3 displays several of these genes along with their
purported functional roles in the regulatory apparatus of
the cell (where known). In this scheme, several
protooncogenes (possibly erbB, ras and sis) may play key
roles in the receptor-mediated activation of so-called
"second messenger" molecules resulting in the activation
of protein kinase C and subsequent multiple cellular
effects, including effects on gene expression. It is not
hard to see how the over expression of these gene's
products or the production of an altered gene product may
effectively "short-circuit" critical regulatory pathways
within the cell resulting in further phenotypic changes.

Applications. The insights into the fundamental
mechanisms of carcinogenesis which oncogene research is
providing will no doubt eventually have a major impact
upon the carcinogenic risk assessment and regulation of
agrochemicals. An example of how this information will
be integrated into this process is in the interspecies
extrapolation of rodent bioassay data to humans by
helping to: 1) determine the appropriateness of the
animal models used in bioassays; and 2) by defining just
how many molecular "events" are required to transform a
normal cell into a fully neoplastic cell. This latter
information, apart from helping to explain on a molecular
level the various stages in this transformation which
have been observed experimentally, may also influence the
choice of mathematical models which are used to estimate
risk based upon animal bioassay results for genotoxic
chemicals. This type of mathematical extrapolation does
not lend itself well for use with nongenotoxic
carcinogens (see below).

A central issue to carcinogenic risk assessment revolves
around the ability to identify animal carcinogens and the
interpretation of this data in terms of human risk. The
vast majority of data on which decisions are made to
protect the public from undue exposure to potential
carcinogens are generated from animal bioassays.
However, these assays are frequently difficult to
interpret, especially when animals exhibiting a high
spontaneous background tumor incidence are used. The use
of such bioassay data in human risk assessment is often
controversial (17-19). While considerable effort has
been expended in evaluating the role of oncogenes in
human cancer, characterizing the presence or absence of
cellular oncogenes in animals used for chronic
carcinogenicity testing has only recently been undertaken
(20-22). One of the most extensively used test species
in oncogenicity bioassays is the hybrid B6C3F1 mouse
which has an average spontaneous liver tumor incidence in

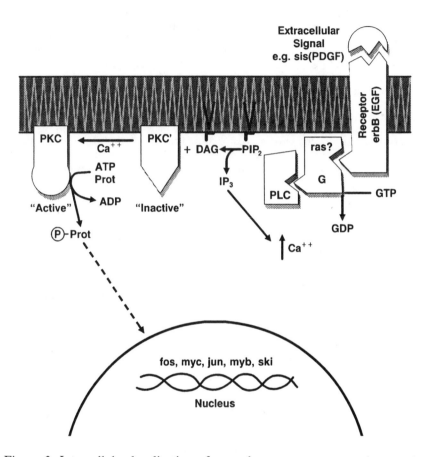

Figure 3. Intracellular localization of several protooncogene products and the involvement of at least some of these in the protein kinase C (PKC) activation by diacylglycerol (DAG) second messengers. (Abbreviations: PDGF, protein derived growth factor; PIP2, phosphotidylinositol bis-phosphate; IP3, inositol triphosphate; G, "G-protein"; PLC, phospholipase C; EGF, epidermal growth factor; GTP, GDP, ATP, and ADP, nucleoside di- and triphosphates; sis, erbB, ras, fos, myc, jun, myb, ski are protooncogenes.) Data are from ref. 16.

males of over 30% (23). Transfection assays of DNA
isolated from spontaneously occurring hepatic tumor
tissue from this strain has revealed the presence of an
active H-ras oncogene in approximately 80% of the animals
examined. DNA isolated from nontumorigenic liver tissue
of these animals was not active in the assay. Specific
mutations that occur in this oncogene isolated from the
tumor tissues of untreated (spontaneous tumors) and
genotoxin treated B6C3F1 mice have been reported to occur
primarily in the 61st codon (22, 24). As shown in Table
4 these changes have involved the transition or
transversion of several bases within this codon. It has
been suggested that the correlation of specific mutations
or patterns of mutations within the codon for spontaneous
and chemically-induced tumor tissue may provide the
ultimate, in vivo, means of identifying the nature of the
chemical-DNA interaction (genotoxic vs nongenotoxic)
(24). However, such an analysis may not differentiate
between the direct or indirect/secondary effects of
chemicals upon DNA. Further research will no doubt
clarify these relationships.

Table 4. The occurrence of mutations within the 61st
codon (CAA) of the H-ras oncogene in the B6C3F1 mouse
(adapted from Wiseman et al. (22); Reynolds et al.(26))

Treatment	# H-ras mutation/# exam.	AAA	CTA	CGA
Spontaneous	15/15	9	3	3
N-hydroxy-2-acetoaminofluorene	7/7	7	0	0
Vinylcarbamate	7/7	0	6	1
1-Hydroxy-2,3-dihydroestragole	10/10	0	5	5
Furan[a]	5/9	4	0	1
Furfural[a]	6/9	5	0	1

[a]Mutations have also been observed in the 13th and
117th codons for these compounds.

The B6C3F1 mouse has also been shown to carry a gene
inherited from its C3H/HeJ parent which is responsible
for an enhanced sensitivity to chemical carcinogens
relative to more resistant strains of mice, even given
the same level of DNA alkylation and repair (25, 26).
Called Hcs for hepatocarcinogen sensitivity, this gene

appears to have a profound impact upon the promotional phase of hepatocarcinogenesis by increasing the proliferation rate of both normal and preneoplastic hepatocytes. Obviously, the interpretation of bioassay data from B6C3F1 mice relative to human risk assessment is confounded by the expression of this gene.

The mathematical modelling of animal bioassay data for the purpose of extrapolating results to lower dose levels, usually beyond experimental verification, is an often controversial yet commonly used tool in quantitative risk assessment procedures. The most appropriate model utilized to estimate risk is based, at least in part, upon an assumption about how many "events" are required to cause a cell to become cancerous. These "events" are generally regarded to be permanent in nature, likely mutations. Two models of some notoriety are the one-hit and the multi-stage linear models which, as the names imply, differ primarily in their basic assumption of how many chemically induced changes (hits) are responsible for tumorigenesis. The multistage models of Armitage and Doll (27), Moolgavkar and Knudson (11) and others provided substantial epidemiological support for the involvement of at least two irreversible steps in a variety of human cancers. Subsequent oncogene research has also provided support for the necessity of multiple changes in the transformation of a normal cell into a fully neoplastic cell at the more basic, cellular level. Several investigators have demonstrated that at least two oncogenes are required to transform primary rodent cultures to a neoplastic phenotype (28-30). However, it is important to note that much work remains before this issue is settled as illustrated by the results of a recent study using transgenic mice which suggested that the number of atlerations required for transformation may be tissue specific (31).

GENOTOXIC vs NONGENOTOXIC CARCINOGENS

Categorizing Carcinogens. The discovery of oncogenes appears to have established the likely "target" sites for mutation within the genome of a cell necessary to initiate the carcinogenic process. However, as noted, not all chemical carcinogens appear to interact to any significant degree with DNA and are not mutagenic in short-term assays which are very sensitive to this interaction. Thus, there appears to be more than one way to get a mutational change in, or a deregulation of, a proto-oncogene (i.e. there's more than one way to get from point A to point B). The practical application of this fact has led to the establishment of two broad categories of carcinogens; genotoxic carcinogens, which are mutagenic or clastogenic or both, and nongenotoxic

carcinogens, which do not interact with the cellular
genome directly. Based upon a chemicals in vitro
mutagenicity and ability to induce DNA repair, Weisburger
and Williams (32) proposed that carcinogens be
categorized as genotoxic (direct acting or primary
carcinogens, procarcinogens, inorganic carcinogens) or
epigenetic (solid-state carcinogens, hormones,
immunosuppressive agents, promotors and cocarcinogens).
The terms epigenetic and nongenotoxic have often been
used interchangeably.

Despite the fact that they may ultimately cause the same
changes at the molecular level of the oncogene, the
distinction between chemicals causing tumors via
primarily a genetic mechanism and those causing tumors
via primarily a nongenotoxic mechanism is of great
significance in carcinogenic risk assessment. Genotoxic
chemicals appear to alter the expression of cellular
oncogenes via a direct interaction with DNA, possibly
causing irreversible initiation of target cells even at
less than toxic dosages. Several natural and synthetic
examples of these compounds are: the mycotoxin aflatoxin
B1, a common contaminant of grains, nutmeats and cotton
seed meal and oil; the grain fumigant methyl bromide; and
the soil fumigant ethylene dibromide. In contrast, the
physiological and biochemical adaptations and/or chronic
cytotoxicity which are believed to be responsible for the
tumorigenic action of many nongenotoxic chemicals display
characteristics of clear thresholds and are often
reversible in nature. This may be true even when
nongenotoxic chemicals are found to bind to
macromolecules since the primary molecular targets of
these compounds are the numerous nonspecific molecules of
the cell, of which there are multiple copies, rather than
DNA. Examples of nongenotoxic animal tumorigens are;
ortho-phenylphenol (OPP), a fungicide; dieldrin, an
organochlorine insecticide; and lactofen, a new
nitrodiphenylether herbicide.

It has become increasingly evident that both genotoxic
and nongenotoxic compounds alike may influence all stages
of the cancer process. However, because of the specific
nature of their chemical and/or pharmacologic action they
have a greater probability of involvement with some
stages relative to others. Genotoxic carcinogens have
their greatest impact on initiation and progression due
to their ability to induce mutational damage to the
genetic material. At cytotoxic dose levels, however,
these compounds will also produce cellular toxicity and
act as a promotional stimulus by inducing regenerative
cell proliferation. In contrast, nongenotoxic
carcinogens primarily exert their influence at the
promotional stage of carcinogenesis. This may occur via
increased cell proliferation or possibly by disruptions

of normal transmembrane and/or intercellular
communication. At high enough dosages these agents may
also influence initiation and progression by indirectly
damaging DNA, for example via increases in reactive
oxygen species following chemically-induced proliferation
of peroxisomes. It is also important to note that the
definition between these various mechanistic
classifications of chemical carcinogenesis may not always
be distinct and may be dose-dependent.

Spontaneous Mutations. Central to nongenotoxic
mechanisms of chemical carcinogenesis is the occurrence
of a finite number of so-called "spontaneous" or
"naturally" occurring mutational events within the genome
of a target cell. These mutations represent the normal
background incidence of mutational change seen in
biological systems. Given the complexity of the cellular
processes which maintain homeostasis, it is not
surprising that there appears to be numerous means, in
the absence of direct DNA alkylation by a synthetic
exogenous compound, by which this may occur. Spontaneous
mutations may originate in one of several ways, among
them; DNA thermodynamic degradation, "naturally"
occurring endogenous or exogenous genotoxic agents, and
replication errors (see reviews by Stott and Watanabe
(33) and Saul and Ames (34)). The DNA molecule and its
maintenance does not represent a static situation in vivo
in the absence of a measurable challenge with a synthetic
genotoxicant, rather it is a dynamic structure with
constant degradation and repair. Such naturally
occurring genotoxic challenges as cosmic radiation, UV
radiation, endogenously produced compounds (e.g.,
formaldehyde), food and airborne exogenous compounds and
loss of bases due to thermodynamic instability and
oxidative damage require a constant DNA surveillance and
repair process. The potential also exists for the
spontaneous alkylation of bases by endogenous S-
adenosylmethionine (35). Degradation due to
thermodynamic instability alone has been estimated to
account for over 11,000 depurinations and several hundred
depyrimidinations and base deaminations of DNA molecules
in a mammalian cell per day (36-38). Deamination of 5-
methylcytosine to form thymidine in particular may result
in some loss of gene regulatory control associated with
this base (39). Based upon the analysis of oxidative
products in the urine of humans, an additional few
thousand lesions/cell/day have been attributed to the
oxidative degradation of DNA (40, 44).

As reviewed by Saul and Ames (34), the sum of the various
DNA degradative "events" could number as high as
200,000/human cell/day. Should these lesions be
improperly repaired prior to replication, a base

transversion or transition may occur and the lesion will
be "fixed" as part of the inherited genetic message.
Indeed, it has been demonstrated in vitro that DNA
polymerases may copy past apurinic sites, and that the
fidelity of DNA synthesis using an apurinic DNA template
may be decreased proportionately to the degree of
depurination (42). In addition, so-called proofreading
repair enzyme(s) (exonucleases) do not appear to
recognize and remove the misincorporated bases at these
apurinic sites (42). A low frequency of DNA base errors
may also result during DNA replication or repair from
base mispairing due to base keto-enol and amino-imino
tautomerization, and anti-syn isomerization about the
glycosyl bonds (43). These base alterations may affect
bond-forming capabilities and characteristics of the base
pairs.

The replication process itself may also introduce errors
into a growing DNA molecule or repair site. As reviewed
by Loeb et al. (44) and Hartman (45), base mismatches may
arise from DNA polymerase base selection errors and
errors by proofreading enzymes due to lack of recognition
or misincorporation. Estimates of in vivo
misincorporation during DNA replication which escape
proofreading repair range from 10E-11 to 10E-8 per base
pair synthesized (44-46). When these sources of DNA base
error and possible mutation associated with the cell
replication machinery are taken together with all other
possible "natural" sources of mutation (above), the end
result in humans appears to be an observable mutation
rate of approximately 10E-6 to 10E-5 mutations per gene
per generation (47, 48). Indeed, it has been estimated
that 10% of all human gametes contain at least one new
mutation of their own as well as several inherited
mutations (49).

Thus, it does appear that spontaneous mutations may occur
in the absence of a measurable synthetic exogenous
genotoxic challenge. Conceivably, some of these
mutations may occur in critical regulatory sites (i.e.
proto-oncogenes) which may ultimately result in
spontaneous cellular transformation. The occurrence of
these spontaneous genetic lesions play a critical role in
the so-called cytotoxic and promotional nongenotoxic
mechanisms of carcinogenicity.

Effect of Cytotoxicity. Tissue degenerative/regenerative
changes in response to a cytotoxic dosage of a chemical
are expected to have a profound impact upon the incidence
and rate at which spontaneous mutations occur and the
subsequent promotion of initiated cells. A shortened
cell cycle in regenerating tissue would leave less time
for DNA repair mechanisms to eliminate misincorporated

bases, apurinic or apyrimidinic sites, and altered bases (pre- and post-replication). Berman et al. (50) have reported an increased mutation frequency (five-fold) in actively dividing versus nondividing rat liver epithelial cells exposed to the mutagenic chemical methyl methanesulfonate suggesting that DNA repair mechanisms in the dividing cells had less time to repair altered DNA bases before the DNA replicated. The result was in a higher mutation rate in the actively dividing cells. Work by Maher et al. (51) on the effect of UV irradiation on human fibroblasts has shown that cell survival is lower and mutation rates higher upon UV irradiation in DNA excision-repair-incompetent cells than in normal cells. Normal cells were also observed to survive a usually lethal and mutagenic UV dose upon being held in confluence (nondividing) for a period of time prior to assessment of survival and mutation rate. These results suggest that this recovery was a result of the increased time given DNA repair mechanisms to remove thymine dimers prior to replication. These workers have also demonstrated that the actual number of DNA alkylation sites removed from the DNA of benzo(a)pyrene exposed fibroblasts by repair mechanisms is directly related to cell survival (52).

In addition to the above considerations, there is an experimental basis for the role of enhanced cellular division and DNA synthesis in carcinogenesis. Tumors frequently develop in chronically inflamed or scarred tissues. Examples in humans include the development of colon cancer in patients suffering from chronic colitis, the development of skin cancer in burn scars, and the association of chronic cirrhosis of the liver with liver tumor development (53-55). Repeated tissue damage with a physical agent (dry ice) and resultant tissue regeneration has been observed to induce skin tumors in mice (56). Similarly, the repeated subcutaneous injection of nonreactive compounds such as glucose, saline, and distilled water have been observed to induce tumors at the site of injection (57). Physical trauma such as partial hepatectomy has also been shown to enhance the tumorigenic effect of thioacetamide, dimethylnitrosamine, and diethylnitrosamine (58, 59). Finally, it has been reported by Slaga et al. (60) and Weeks et al. (61) that the promotion of dimethylbenzanthracene-induced skin tumors by the phorbol esters may be inhibited by anti-inflammatory steroids and retinoic acids which inhibit DNA synthesis, cell proliferation, and inflammation.

Cytotoxicity accompanied by compensatory cellular division are frequently observed in chronic animal bioassays where the emphasis is placed upon attaining a "maximum tolerated dosage" in an attempt to increase the

sensitivity of these assays by utilizing the highest
possible dose levels of a test chemical. Consequently,
induction of cytotoxicity in target organs may have a
profound "cocarcinogenic" effect with low levels of
endogenous carcinogens and a potent promotional effect
upon spontaneously initiated cells. This fact takes on
added significance as most animals used in bioassays have
an appreciable rate of spontaneous tumor development in
the absence of any synthetic xenobiotic treatment (e.g.
lung and liver tumors in the B6C3F1 mouse; leukemia and
testicular cancer in the Fischer 344 rat; and mammary
tumors in the Sprague-Dawley rat). Thus, nongenotoxic
compounds could increase tumor yield without themselves
"initiating" any new tumors. Further, cytotoxic dose
levels of genotoxic chemicals could be expected to cause
an additional, disproportionate, increase in the
carcinogenic potency of these compounds with increasing
dose, by expanding the clonal population of altered
cells.

To visualize the effects that cytotoxic dosages of
chemicals may have on the tumorigenic process, a computer
model derived from the work of Gehring and Blau (4) and
Reitz and Watanabe (62) was developed by Reitz (63)
(Figure 4). This model contains a series of simultaneous
differential equations that were evaluated with ACSL
(Mitchell and Gauthier Associates, Concord, MA) to
calculate the number of cells containing two critical
site mutations resulting in the activation of one or more
proto-oncogenes. This simulation allows for the
depletion of critical macromolecules through reaction
with activated chemical (genotoxic or nongenotoxic) as
governed by a homeostatic mechanism such as those known
to control levels of cellular components; the saturable
synthesis of the critical macromolecule; a cellular
replication rate which was linked to the concentration of
the critical macromolecule, assuming it has an important
function in this process; and detoxification and repair
rates. The critical molecule may be one of numerous
molecules necessary for cellular homeostasis (e.g.
proteins, mRNA, GSH, etc.). Sufficient depletion or
alteration of these molecules results in cell death and
the subsequent division of surviving cells to replace
those lost. The greater the rate of cell turnover, the
greater the rate of fixation of chemically-induced or
spontaneously-occurring DNA lesions as heritable
mutations. The rate of formation of mutations may thus
increase disproportionately with increasing dose at toxic
levels of either genotoxic or nongenotoxic compounds, but
especially in the case of the former with their added
ability to alter DNA directly.

Figure 5 presents a simulation of the relative number of
potentially neoplastic cells which would be expected to

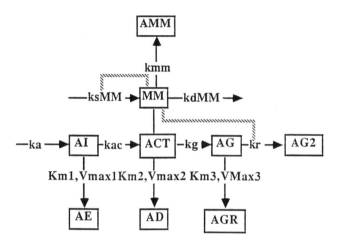

Figure 4. Diagram of the computer model used to simulate the effects of cytotoxicity on the carcinogenic process. Abbreviations: prefix "k" defines the rate constants for absorption (a), metabolic activation (ac), binding to genetic material (g), synthesis of MM (sMM), degradation of MM (dMM), binding to MM (mm); prefix "A" defines amount of chemical absorbed (I), eliminated (E), activated (CT), detoxified (D), bound to MM (MM), bound to genetic material (G), AG repaired (GR), and DNA damage consolidated by replication (G2); MM denotes concentration of the critical macromolecule necessary for cell survival. The amounts of AE, AD and AGR are determined by saturable (Michaelis-Menten) processes (denoted by Km and Vmax).

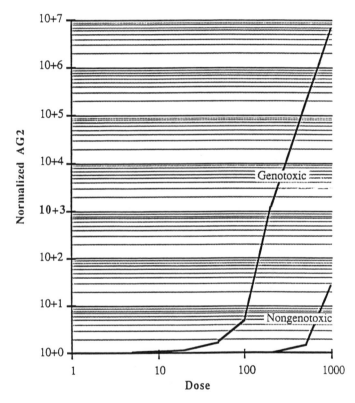

Figure 5.   Simulated values of the amount of replicated
genetic lesions (AG2) for nongenotoxic and genotoxic
agents as a function of dose.

occur over a thousand-fold dose range of a relatively nongenotoxic chemical and a potent genotoxic chemical. In this simulation it was assumed that the homeostatic mechanism controlling macromolecular synthesis prevented significant depletion of these molecules at dose units of ≤ 100. However, saturation of detoxification and elimination mechanisms following a dose of 1000 units results in a transient depletion of the macromolecule and the death of a certain percentage of the cell population; replaced, in turn, by cellular regeneration. The model predicts that a series of cytotoxic doses of a nongenotoxic agent would be expected to produce the same effect as a series of doses of a weakly genotoxic agent as a result of the endogenous processes providing a low background of alterations in DNA. What is also striking about the results of this exercise is the nonlinearity of the response obtained at higher dose levels, nearly seven orders-of-magnitude higher than that predicted based upon a linear extrapolation of the response obtained at 1 dose unit. This dramatic increase results from the fact that cellular replication, and thus consolidation of macromolecular changes, has increased at the very time that DNA repair mechanisms are saturated and/or provided less time to function and hence less efficient in removing DNA damage.

While only a simulation, this model still serves to illustrate: 1) the potential impact of cytotoxicity upon the dynamics of DNA turnover and the background rate of spontaneous mutation/proto-oncogene activation or, in the case of genotoxins, the rate of chemically-induced mutation; and 2) the potentially large differences in potency, at a macromolecular level, which may exist between these two classes of carcinogens. One factor which this model does not take into account, however, is the potential impact of purely receptor-mediated promotional activity of a cytotoxic compound.

The activation of protein kinase C and subsequent induction of cell proliferation by cytotoxic chemicals has been suggested by Roghani et al. (64). The inappropriate activation of this enzyme by a chemical-receptor complex is believed to play a role in the promotion process. The activation of protein kinase C may subsequently result in the activation of a number of cytoplasmic proteins through inductional phosphorylation and set off a chain of events which includes enhanced cell proliferation in the absence of regenerative cell replication. The distinction between cytotoxicity and promotion may not always be clear. As recently discussed by Trump and Berezesky (65), chemically-induced alterations in the cellular dynamics of calcium regulation may result in a wide range of changes, many of which are similar to those caused by protein kinase C

activation. However, more research is obviously needed
before any firm conclusions can be drawn regarding the
interrelationship of cytotoxicity and protein kinase C-
mediated promotion.

An example of an agrochemical which appears to cause
tumors in rodents via chronic regenerative DNA synthesis
in response to cytotoxicity is ortho-phenylphenol (OPP)
and its sodium salt (SOPP). These compounds are used as
fungicides and antibacterial agents for the post-harvest
treatment of fruits and vegetables (66). Hiraga and
Fujii (67, 68) reported that Fischer 344 rats consuming
diets with >1% SOPP or OPP (approximately 700 mg/kg/day)
developed tumors of the urinary tract, especially bladder
tumors. Yet, as shown in Table 5, OPP has been found to
give a negative response when tested in an impressive
number of assays designed to detect any ability of this
molecule or its metabolites to interact directly with a
target prokaryote or mammalian cell genome. In addition,
Reitz et al. (69) could detect no covalent binding of
radiolabelled SOPP to the DNA of bladder tissue of rats
administered a relatively high dose level of this
compound (Table 5). The weight of the evidence clearly
indicates a lack of genotoxicity of OPP and its sodium
salt.

TABLE 5. Results of assays for potential genotoxicity
of OPP and SOPP (66, 69, 70)

ASSAY	RESULT
B. subtilis H17 and M45	
rec-assay	−
E. coli WP2 hcr + and −	
metabolic activation	−
S. typhimurium TA1535, 1537,	
1538, 98, and 100 (with or	
without metabolic activation)	−
S. typhimurium G46 host mediated	
in mice	−
Primary rat hepatocyte UDS	−
Bone marrow cytogenetics in rats	−
	+
Chromosomal aberration in Chinese	
Hamster Ovary cells	−
Dominant-lethal and specific locus	
in mice	−
SCE in Chinese Hamster Ovary cells	−
In vivo covalent binding to DNA	
(rat urinary bladder)	−
(sensitivity = 1 alkylation/10E6 nucleosides)	

Reitz et al. (69) studied the metabolism of SOPP in F344 rats and found a pronounced dose-dependency in the metabolic pathways. At low doses (3.3-33 mg/kg) most of the SOPP administered by gavage was eliminated as glucuronide or sulfate conjugates of SOPP. However, when a higher dose of SOPP (326 mg/kg) was administered, 25-30% of the urinary radioactivity was recovered in the form of more highly oxidized species. These hydroquinones are hypothesized to be responsible for the toxicity of OPP.

Reitz et al. (71) subsequently found that in vivo macromolecular binding of 14C-SOPP increased disproportionately as the primary metabolic pathways were saturated (Table 6) and suggested that the toxicity of SOPP was related to production of the more highly oxidized metabolite of SOPP. Rates of cellular division in bladder epithelial cells (visualized by microautoradiography) increased 50-70 fold after administration of a high dose of 326 mg/kg SOPP (62). Therefore, only at high metabolically saturating doses does toxicity, hyperplasia and tumorigenicity occur.

Table 6. In vivo macromolecular binding in tissues of male Fischer 344 rats administered ^{14}C-SOPP

DOSE (mg/kg/day)	Liver	Kidney	Bladder
36	3.7± .6	4.9± .5	1.9
73	9.0± .6	8.8± 2.4	2.4
144	28.0± 11[a]	37.0± 5.5	24.0
365	485.0± 210[a]	200.0± 22[a]	400.0[a]

[a]Indicates significant deviation from linearity by regression analysis.

Physiological Adaptation. The tumorigenic activity of numerous nongenotoxic chemical carcinogens, including several herbicides and insecticides, has been found to be closely associated with very distinct biochemical and physiological adaptations which occur in response to the administered chemical. Two major classes of these compounds have emerged; those which induce the proliferation of the subcellular organelle, the peroxisome, and those inducing the activity of the cytochrome P-450's dependent mixed function oxygenases of the smooth endoplasmic reticulum (SER). Some members of the latter group of compounds exhibit classical

promotional activity which appears to govern their
tumorigenic activity (i.e. so-called "pure promotors").

Peroxisomes are single membrane bound organelles which
contain a variety of primarily catabolic enzymes.
Notable among them are several oxidases which generate
hydrogen peroxide as a metabolic byproduct and catalase,
which degrades this reactive compound. A diverse group
of chemicals have been found to cause the proliferation
of peroxisomes in rodents and induce the activities of
peroxisomal enzymes, sometimes as much as 10 to 20-fold.
Significant among these is the induction of the activity
of peroxisomal fatty acid beta-oxidation, a hydrogen
peroxide generating enzyme system, which may far outstrip
the induction of the activity of peroxisomal çatalase,
the enzyme responsible for the degradation of the
peroxide formed. This relationship, the strong
qualitative statistical relationship between peroxisome
proliferation and tumorigenic activity in rodent
bioassays, and the nongenotoxicity of these chemicals
have led J. Reddy and coworkers to propose that
peroxisome proliferators form a "novel class of chemical
carcinogens" (72).

Several important characteristics of chemicals which
induce the proliferation of peroxisomes in animals is
that they display clear thresholds and species-dependency
in the induction of proliferation (see review by Stott
(73) and references contained therein). In rodent
bioassays, no tumorigenic response is observed at
nonproliferating dose levels and higher mammalian species
are quite refractory to the proliferative effects of
these chemicals. Indeed primates, including humans,
appear to be so refractory to the chemically-induced
proliferation of peroxisomes that this response, and any
carcinogenic hazard associated with it, is generally
considered to represent a rodent specific phenomenon.
The reason for this pronounced species dependency does
not appear to be solely related to species differences in
the pharmacokinetics of the administered proliferator.
Rather, species differences have been speculated to be
related to differences in the concentration of an, as yet
unidentified, specific cell receptor(s) and/or to an
"initiating" metabolic perturbation such as depressed
mitochondrial function resulting in substrate "overload"
and subsequent substrate induction of peroxisome
proliferation.

A schematic representation of the peroxisomal
proliferation theory of chemical carcinogenesis in
rodents is presented in Figure 6. In this model, the
chronic induction of the activity of a normally occurring
peroxisomal enzyme system is believed to greatly
accelerate the process of cellular oxidative damage.

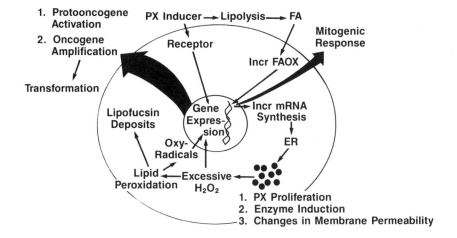

Figure 6. Schematic showing the theory of peroxisome-mediated mechanism of carcinogenesis (PX, peroxisome; ER, endoplasmic reticulum; FA, fatty acid; FAOX, peroxisomal and mitochondrial FA beta-oxidation activity) (modified from (74)). In this scheme; 1) the proliferation of peroxisomes is initiated by a receptor-mediated or induced substrate-overload mechanism, 2) increased expression of structural genes results in de novo synthesis of peroxisomes and induction of FAOX enzymes (induction of P-452 dependent omega-oxidation of FA not shown), 3) The resultant over production of hydrogen peroxide results in membrane degradation, accumulation of lipofuscin and an increase in oxidative stress, 4) hydrogen peroxide or oxidative products alter gene expression resulting in the activation of a protooncogene which may ultimately result in cellular transformation and tumor development.

Initially, an effective dosage of a peroxisome
proliferator, via a receptor or substrate   mediated
mechanism (shown as a result of increased lipolysis),
causes an increased expression of peroxisomal structural
genes resulting in the de novo synthesis of peroxisomes
and a disproportionate synthesis of peroxisomal enzymes.
Upon repeated administration of the compound, the
disproportionate increase in the activities of the fatty
acid beta-oxidation pathway and possibly other oxidases
relative to the catalase activity of these peroxisomes
results in a significant elevation in peroxisomal levels
of hydrogen peroxide. Hydrogen peroxide itself, or
hydroxyl radicals formed by its iron-catalyzed
degradation, may subsequently escape elimination by
defense mechanisms and react with peroxisomal or other
subcellular organelle membrane lipids to initiate the
auto-oxidation of fatty acids. This may result in the
generation of a variety of reactive oxygen species and
auto-oxidation products (e.g., aldehydes, polymers,
organoperoxides and conjugated dienes). Some of these
compounds and/or hydrogen peroxide itself are believed to
interact with the genome of the cell resulting in DNA
damage and the possible anaplastic transformation of the
cell.

While the exact molecular mechanism of inducing
hepatocellular tumors in rodents administered peroxisome
proliferators has yet to be elucidated, it is speculated
that all stages of the carcinogenesis process may be
involved. Any mitogenic effects or prolonged
regenerative DNA synthesis resulting from either a direct
effect of an administered peroxisome proliferator or as a
result of peroxisome proliferation would be expected to
especially influence the latter stages of cellular
transformation of these cells, possibly enhancing their
tumorigenic potency. In this regard, it is important to
note that several investigators have reported that
mitogenic potency correlates well with the tumorigenic
potential of peroxisome proliferating agents (75, 76).
Still others have suggested a significant involvement of
chronic regenerative DNA synthesis in this process (77-
79).

An example of an agrochemical which has been shown to
cause tumors in rodents via a peroxisome proliferation
mechanism of action is the broad spectrum diphenyl ether
herbicide, lactofen (80, 81). Lactofen has been shown to
induce an increased incidence of liver tumors in rats and
mice following chronic administration in the absence of
observable genotoxicity as determined in numerous short-
term assays. As shown in Table 7, tumorigenic levels of
lactofen were observed to cause changes in liver weights,
histology, peroxisome numbers and peroxisomal enzyme
activities characteristic of known peroxisome

proliferators, in this case the positive control, the potent peroxisome proliferator nafenopin (80). Consistent with the extreme species-dependency of the activity of other peroxisome proliferative agents, lactofen failed to induce the proliferation of hepatocellular peroxisomes in chimpanzees administered comparable dosages as those causing pronounced proliferation in rats (81).

Table 7. Effect of a carcinogenic dose level of lactofen on hepatic parameters in male CD-1 mice (adapted from Butler et al. (80))

Group	Liver Wt (g/100g BW)	ENZYME ACTIVITY (Units/mg Prot.) Catalase[a]	FAOX[b]	Histo-pathology	PX:Mito[c] Ratio
Control	3.92 ±.35	343 ±160	3.07 ±.73	–	1:4.8
Lactofen	7.67* ±.36	939* ±250	40.0* ±11	Hypertrophy	1:1.5
Nafenopin (+control)	12.8* ±.79	695* ±210	48.7* ±23	Hypertrophy	ND

[a]U=umol. H2O2 hydrolyzed/min.
[b]Peroxisomal beta-oxidation of fatty acids (U=umol. dichlorofluoroscein/min).
[c]Peroxisome:mitochondria ratio
*Statistically significant difference relative to controls (p<0.005).

Another apparent nongenotoxic mechanism of carcinogenesis has been identified by the association of the induction of classic hepatic drug metabolizing enzymes and the formation of liver tumors in rodents. This group of chemicals encompasses numerous pharmaceutical and industrial chemicals (e.g..phenobarbital, PCBs) and an important class of agrochemicals, the organochlorine insecticides. As reviewed by Tennekes (82) these compounds have been reported to cause tumors in a variety of rodent bioassays yet display little or no genotoxic activity in short-term and animal assays of mutagenicity or clastogenicity. Furthermore, the enzyme induction and associated liver hypertrophy which occur in treated animals are clearly reversible and have well defined thresholds. It was concluded by Tennekes (82) that this group of compounds likely induced an increased incidence of tumors in susceptible animal models (i.e. mouse strains having a relatively high spontaneous incidence of

liver tumors) via the promotion of existing
"spontaneously" occurring initiated cells. Consistent
with this, DDT, PCBs and dieldrin have been shown to
promote the carcinogenic effects of known genotoxic
agents in numerous initiation-promotion bioassays much as
phenobarbital does (83-86). Should the organochlorines
function much as phenobarbital (i.e. as a classical tumor
promotor), than it is possible that these compounds may
also influence the activity of protein kinase C. The
activation of this enzyme may in turn result in a great
number of cellular alterations including enhanced cell
replication, a significant factor in the activity of a
tumor promotor. Consistent with this has been the
observation by Busser and Lutz (87) that phenobarbital
and several organochlorine insecticides (aldrin, DDT)
cause a roughly 3-fold increase in DNA synthesis at non-
necrotizing dose levels in species and strains of rodents
which were sensitive to the tumorigenic effects of these
compounds. No induction of DNA synthesis was observed in
resistant species or strains of rats or mice at dose
levels tumorigenic in sensitive animals.

An example of this class of nongenotoxic carcinogens is
the organochlorine insecticide dieldrin. Dieldrin has
been shown to cause a significant increase in the
incidence of liver tumors in a variety of mouse strains,
all of which routinely display a relatively high rate of
spontaneous tumor formation. Yet, as shown in Table 8,
dieldrin does not appear to be genotoxic as measured by
several bacterial and mammalian assays nor does it
covalently bind DNA in vivo. This compound causes
chronic hepatic hypertrophy in treated animals; however,
unlike peroxisome proliferative agents, this effect
appears to be related to the proliferation of SER and the
induction of the activities of cytochrome P-450's
dependent mixed function oxygenases housed within the

Table 8.   Short-term genotoxicity assays of dieldrin
(82)

Assay	Results
S. typhimurium Mutagenicity	−
Mouse Host Mediated Bacterial Mutagenicity	−
Mouse Dominant Lethal	−
Rat, Mouse DNA Strand Damage	−
DNA Covalent Binding	rat-1.52/10E9 nucleotides mouse- 0.13 to 0.58/10E9 nucleotides

SER. For example, the relative liver weights were elevated approximately 50% following 15 and 52 weeks of administration of 10 mg/kg/day dieldrin via the diet. This increase in liver weight correlated to a roughly 9-fold increase in the specific activity of hepatic p-nitroanisole O-demethylase activity at both time points.

THE FUTURE: A BIOLOGICALLY-BASED APPROACH TO CHEMICAL CARCINOGENESIS

As demonstrated by the above discussion and examples, chemical carcinogenesis is a complex, integrated process involving many highly interdependent events; the metabolism and pharmacokinetics of the chemical which define the exposure of the target tissue to the active molecule, effects upon cellular and macromolecular dynamics, and proto-oncogene activation. A comprehensive model which encompasses these three areas has been proposed by Conolly et al. (88) is displayed in Figure 7. This model essentially embodies at the cellular/molecular level the definition of carcinogenic hazard, the interaction of exposure and toxicity, at the macroscopic whole body level. It is composed of three basic sections: 1) a physiologically-based pharmacokinetic portion to determine the rate of exposure of a target cell to the active molecule following exposure of the whole animal; 2) a portion designed to estimate the effects of differing biological mechanisms of action (genotoxic or nongenotoxic or any mixture of the two); and 3) a two-stage cancer model in which cells may be i) normal, ii) carrying one mutation in a critical site (i.e. in a proto-oncogene) or iii) malignant (carrying two mutations). While still in its infancy and with validation experiments still underway, this model represents a biologically-based,more wholistic, approach to quantifying the effects of chemical carcinogens upon animals and provides the framework necessary for the dose-related, route-related, interspecies extrapolation of carcinogenic risk.

CONCLUSION

This paper has examined the theoretical and experimental basis for the existence of different mechanisms of carcinogenic activity of chemicals within the context of the basic definition of what constitutes a carcinogenic hazard. Examined in this way, the mechanism of carcinogenic action of a chemical may encompass exposure, cell dynamics and gene activation as interrelated parts of the process of oncogenesis. The balance of these effects, or segments, of the mechanistic process define the potential carcinogenic activity of a chemical. Yet, as complex and incompletely understood in toto as this

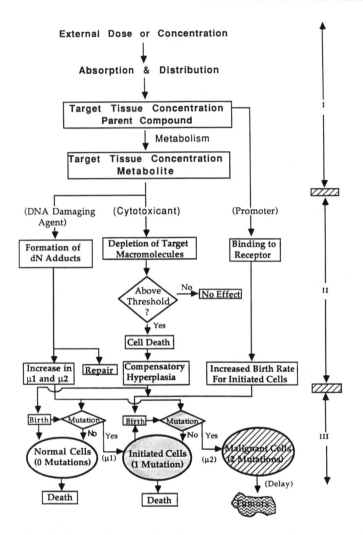

Figure 7.   Schematic of the Pharmacodynamic Cancer Model
(modified from (88)). Segment I is composed of a
physiologically based description of the processes
controlling carcinogen pharmacokinetics which converts
external dose to a realistic target tissue concentration
of the toxicant (parent chemical or metabolite). Segment
II is composed of biochemical mechanisms for genotoxins,
cytotoxins and promotors resulting in mutations or
increased cell turnover or both. Segment III consists of
a two-stage cancer model where cells are either normal (0
critical site mutations), intermediate (1 critical site
mutation) or malignant (2 critical site mutations). The
linkages between segments II and III define how the
different biochemical mechanisms affect the cell growth
and mutation parameters of the cancer model.

process is, knowledge about the likely mechanism of action may still have a very practical and significant application in the ranking of the carcinogenic risk of chemicals. Armed with the results of assays of genotoxicity, cytotoxicity, and specific physiological adaptations (e.g., peroxisome proliferation), it is possible to divide most chemical carcinogens into general categories for which there is overwhelming experimental evidence demonstrating major differences in potential carcinogenic risk.

In contrast to genotoxic carcinogens, nongenotoxins: 1) are dependent upon spontaneous alterations in the genome or the effects secondary to some induced pharmacologic or toxicologic alteration for initiation to occur; 2) usually require the chronic administration of relatively large dosages; 3) are most often associated with clearly identifiable biochemical, physiological and morphological changes; and 4) have clearly defined thresholds for those responses which appear to be necessary for tumor development to occur. Thus, in a practical application of mechanistic data, nongenotoxic carcinogens such as OPP, dieldrin and lactofen are not considered to pose the same carcinogenic threat to animals at lower dosages as the potent, naturally occurring carcinogen aflatoxin B1; even if each compound were observed to induce the same incidence of tumors in a rodent test species at the same relatively high dose level(s) (i.e. have similar potencies). Clearly, only by the judicious application of mechanistic data for chemical carcinogens may a more scientifically sound and defensible risk assessment of animal carcinogens be made.

ACKNOWLEDGMENTS

The authors wish to acknowledge the efforts of Dr. A.M. Schumann for his review of this manuscript and Ms. P. McRoberts in manuscript preparation.

REFERENCES

1. Boveri, T. The Origin of Malignant Tumors; Williams and Watkins: Baltimore, MD, 1929.
2. Topal, M. D. Carcinogenesis 1988, 9, 691-696.
3. Ludlum, D. B.; Papirmeister, B. In Mechanism DNA Damage and Repair; Semic, M.; Grossman, L.; Upton, A. C.; Eds.; Plenum Press: NY, 1986; pp 119-126.
4. Gehring, P. J.; Blau, G. E. J. Environ. Pathol. Toxicol. 1977, 1, 163.
5. Lutz, W. K. Mut. Res. 1979, 65, 289-356.
6. Trosko, J. E.; Chang, C. C. Photochem. Photobiol. 1978, 28, 157-168.

7. Berenblum, I. J. Natl. Cancer Inst. 1978, 60, 723-
    726.
8. Mintz, B.; Illmensee, K. Proc. Natl. Acad. Sci. 1975,
    9, 3585.
9. McKinnell, R. G.; Deggins, B. A.; Labat, D. D.
    Science 1969, 165, 394-396.
10. Papaioannon, V. E.; McBurney, M. W.; Garnder, R. L.;
    Evans, M. J. Nature 1975, 258, 70-73.
11. Moolgavkar, S. H.; Knudson, A. G. J. Natl. Cancer
    Inst. 1981, 66, 1037-1052.
12. Shilo, B. Z.; Weinberg, R. A. Nature 1981, 300, 607-
    609.
13. Reddy, E. P.; Reynolds, R. K.; Santos, E.; Barbacid,
    M. Nature 1982, 300, 142-152.
14. Schwab, M.; Alitalo, K.; Varmus, H. E.; Bishop, D.;
    George, A. Nature 1983, 303, 497-501.
15. Palciani, S.; Santos, E.; Lauver, A. V.; Long, L. K.;
    Robbins, K. C.; Barbacid, M. Proc. Natl. Acad. Sci.
    USA 1982, 79, 2845-2849.
16. Bell, R. M. Cell 1986, 45, 631-632.
17. Interdisciplinary Panel on Carcinogenicity. Science
    1984, 225, 682-686.
18. Task Force of Past Presidents - Society of Toxicology.
    Fundam. Appl. Toxicol. 1982, 2, 101-107.
19. Nutritional Foundation of the International Expert
    Advisory Committee to the Nutrition Foundation, Inc.,
    Washington, D.C., 1983.
20. Fox, T. R.; Watanabe, P. G. Science 1985, 228, 596-
    597.
21. Reynolds, S. H.; Stowers, S. J.; Maronpot, R. R.;
    Anderson, M. W.; Aaronson, S. A. Proc. Natl. Acad.
    Sci. 1986, 83, 33-37.
22. Reynolds, S. H.; Stowers, S. J.; Maronpot, R. R.;
    Aaronson, S. A.; Anderson, M. W. Science 1987, 237,
    1309-1316.
23. Chan, C. NTP Technical Report CAS No. 597-25-1. 1984.
24. Wiseman, R. W.; Stowers, S. J.; Miller, E. C.;
    Anderson, M. W.; Miller, J. A. Proc. Natl. Acad. Sci.
    1986, 83, 5825-5829.
25. Drinkwater, N. R.; Ginsler, J. J. Carcinogenesis
    1986, 7, 1701-1707.
26. Hanigan, M. H.; Kemp, C. J.; Ginsler, J. J.;
    Drinkwater, N. R. Carcinogenesis 1988, 9, 885-890.
27. Armitage, P.; Doll, R. Br. J. Cancer 1957, 11, 161-
    169.
28. Land, H.; Parada, L. F.; Weinberg, R. A. Science
    1983, 222, 771-778.
29. Ruley, H. Nature 1983, 304, 602-606.
30. Newbold, R.; Overell, R. Nature 1983, 304, 648-651.
31. Muller, W. J.; Sinn, E.; Pattengale, P. K.; Wallace,
    R.; Leder, P. Cell 1988, 54, 105-115.

32. Weisburger, J. H.; Williams, G. M. In Toxicology: Basic Science of Poisons; Doull, J.; Klaassen, C. D.; Amdur, M. O.; Eds.; MacMillan: NY, 1980; Vol. 2, pp. 84-138.
33. Stott, W. T.; Watanabe, P. G. Drug Metab. Rev. 1982, 13, 853-873.
34. Saul, R. L.; Ames, B. N. In Mechanisms of DNA Damage and Repair; Simic, M. G.; Grossman, L.; Upton, A. C.; Eds.; Plenum Press: NY, 1986; pp 529-535.
35. Barrows, L. R.; McGee, P. N. Carcinogenesis 1982, 3, 349-351.
36. Lindahl, T.; Nyberg, B. Biochemistry 1972, 11, 3610-3618.
37. Lindahl, T.; Nyberg, B. Biochemistry 1974, 13, 3405-3410.
38. Lindahl, T.; Karlstrom, O. Biochemistry 1973, 12, 5151-5154.
39. Razin, A.; Riggs, A. D. Science 1980, 210, 604-610.
40. Hollstein, M. C.; Brooks, P.; Linn, S.; Ames, B. N. Proc. Natl. Acad. Sci. USA, 1984, 81, 4003-4007.
41. Ames, B. N.; Saul, R. L.; Schwiers, E.; Adelman, R.; Cathcart, R. Proc. Symp. on Molecular Biology of Aging: Gene Stability and Gene Expression, October 19, 1984, NY.
42. Shearman, C. W.; Loeb, L. A. J. Mol. Biol. 1979, 128, 197
43. Topal, M. D.; Fresco, J. R. Nature 1976, 263, 285-289.
44. Loeb, L. A.; Weymouth, L. A.; Kunkel, T. A.; Gopinathan, K. P.; Beckman, R. A.; Dube, D. K. Cold Spring Harbor Symp. Quant. Biol. 1978, 43, 921-927.
45. Hartman, P. E. Environ. Mut. 1980, 2, 3-16.
46. Drake, J. W. Nature 1969, 221, 1132.
47. Neel, J. V. Proc. Natl. Acad. Sci. 1973, 70, 3311-3315.
48. Stevenson, A. C.; Kerr, C. B. Mut. Res. 1967, 4, 339-352.
49. Drake, J. W. In Advances in Modern Toxicology; Flamm, W. G.; Mehlam, M. A.; Eds.; Hemisphere Publishing Corp.: NY, 1978, pp. 9-26.
50. Berman, J. J.; Tong, C.; Williams, G. M. Cancer Lett. 1978, 4, 277-283.
51. Maher, V. M.; Dorney, D. J.; Menrala, A. L.; Konze-Thomas, B.; McCormick, J. Mut. Res. 1979, 62, 311-323.
52. Yang, L. L.; Maher, V. M.; McCormick, J. J. Proc. Natl. Acad. Sci. 1980, 77, 5933-5937.
53. Berenblum, I. Arch. Pathol. 1944, 38, 233.
54. Chan, C. H. Med. Clin. North Am. 1975, 59, 989-994.
55. Laroye, G. J. Lancet 1974, June 1, 1097-1100.
56. Berenblum, I. Br. J. Exp. Pathol. 1929, 10, 179.
57. Grasso, P.; Golberg, L. Fd. Cosmet. Toxicol. 1966, 4, 297-320.

58. Craddock, V. M. In Primary Liver Tumors; Remmer, H.;
Bolt, H. M.; Popper, H. and Bannaschi, P.; Eds.; MIT
Press: Boston, MA, 1978, pp 377-383.
59. Date, P. A.; Gotoskar, S. V.; Bhide, S. V. J. Natl.
Cancer Inst. 1976, 56, 493-497.
60. Slaga, T. J.; Fisher, S. M.; Viaje, A.; Berry, D. L.;
Bracken, W. M.; LeClerc, S.; Miller, D. R. In
Carcinogenesis, Mechanisms of Tumor Promotion and
Carcinogenesis; Slaga, T. J.; Sivak, A.; Boutwell, R.
K.; Eds.; Raven Press: NY, 1978, Vol. 2; pp. 173-195.
61. Weeks, C. E.; Slaga, T. J.; Hennings, H.; Gleason,
G. L.; Bracken, W. M. J. Natl. Cancer Inst. 1979, 63,
401-406.
62. Reitz, R. H.; Watanabe, P. G. In Banbury Report 19:
Risk Quantitation and Regulatory Policy; Hoel, D. G.;
Merrill, R. A.; Perera, F. P.; Eds.; Cold Spring
Harbour, NY; 1985; pp. 241-251.
63. Reitz, R. H.; Watanabe, P. G. In Banbury Report 25:
Nongenotoxic Mechanisms in Carcinogenesis;
Butterworth, B.; Slaga, T.; Eds.; Cold Spring Harbor
Laboratory Press: Cold Spring Harbour, NY, 1987; pp.
107-112.
64. Roghani, M.; DaSilva, C.; Castagna, M. Biochem.
Biophys. Res. Com. 1987, 142, 738-744.
65. Trump, B. F.; Berezesky, I. K. Carcinogenesis 1987,
8, 1027-1031.
66. IARC. IARC Monographs on the Evaluation of the
Carcinogenesis Risk of Chemicals to Humans; 1982, Vol.
30, pp. 329-344.
67. Hiraga, K.; Fujii, T. Fd. Cosmet. Toxicol. 1981, 19,
303-310.
68. Hiraga, K.; Fujii, T. Fd. Cosmet. Toxicol. 1984, 22,
865-870.
69. Reitz, R. H.; Fox, T. R.; Quast, J. F.; Hermann,
E. A.; Watanabe, P. G. Chem. Biol. Interact. 1983,
43, 99-119.
70. Shirasu, Y.; Moriya, M.; Kato, K.; Tezuka, H.; Henmi,
R.; Shingu, A.; Kaneda, M.; Teramoto, S. Mut. Res.
1978, 54, 227 Abstract.
71. Reitz, R. H.; Fox, T. R.; Quast, J. F.; Hermann,
E. A.; Watanabe, P. G. Toxicol. Appl. Pharmacol.
1984, 73, 345-349.
72. Reddy, J. K.; Azarnoff, D. L.; Hignite, C. E. Nature
1980, 283, 397-398.
73. Stott, W. T. Regul. Toxicol. Pharmacol. 1988, 8,
125-159.
74. Reddy, J. K.; Lalwani, N. D. CRC Crit. Rev. Toxicol.
1983, 12, 1-58.
75. Bieri, F.; Muakkassah-Kelly, S.; Waechter, F.;
Staubli, W.; Bentley, P. In Peroxisomes in Biology
and Medicine; Fahimi, H. D.; Sies, H.; Eds.; Springer-
Verlag: NY, 1987, pp. 286-294.

76. Elliott, B. M.; Elcombe, C. R. Carcinogenesis 1987, 8, 1213-1218.
77. Smith-Oliver, T.; Butterworth, B. E. Mut. Res. 1987, 188, 21-28.
78. Conway, J. C.; Tomaszewski, K. E.; Cattley, R. C.; Marsmann, D. S.; Melnick, R. L.; Popp, J. A. The Toxicologist 1988, 8, 664 Abstract.
79. Marsman, D. S.; Cattley, R. C.; Popp, J. A. The Toxicologist 1987, 7, 658 Abstract.
80. Butler, E. G.; Tanaka, T.; Iehida, T.; Maruyama, H.; Leber, A. P.; Williams, G. M. Toxicol. Appl. Pharmacol. 1988, 93, 72-80.
81. Leber, A. P.; Fisher, C.; Couch, R.; Erickson, M.; Butler, E.; Maruyama, H.; Williams, G. The Toxicologist 1988, 8, 706 Abstract.
82. Tennekes, H. A. The Relationship Between Microsomal Enzyme Induction and Liver Tumour Formation; Centre for Agricultural Publishing: Wageningen, Netherlands, 1979.
83. Peraino, C.; Fry, R. J. M.; Staffeldt, E.; Christopher, J. P. Can. Res. 1975, 35, 2884-2890.
84. Peraino, C.; Fry, R. J. M.; Staffeldt, E.; Christopher, J. P. Fund. Cosmet. Toxicol. 1977, 15, 93-96.
85. Kimura, N. T.; Kanematsu, T.; Baba, T. Z. Krebsforsch. 1976, 87, 257.
86. Weisburger, J. H.; Madison, R. M.; Ward, J. M.; Viguera, C.; Weisburger, E. K. J. Natl. Cancer Inst. 1975, 54, 1185-1188.
87. Busser, M-T.; Lutz, W. K. Carcinogenesis 1987, 8, 1433-1437.
88. Conolly, R. B.; Reitz, R. H.; Clewell, H. J.; Anderson, M. E. Toxicol. Lett. 1989, 43, 189-200.

RECEIVED June 28, 1989

Chapter 5

Critical Events and Determinants in Multistage Skin Carcinogenesis

Thomas J. Slaga

University of Texas, M. D. Anderson Cancer Center, Science
Park—Research Division, P.O. Box 389, Smithville, TX 78957

A large number of both natural and synthetic chemicals
have been shown to induce cancer in experimental
animals and more than likely some of these may play a
role in the induction of cancer in man. It is generally
recognized all chemical carcinogens are either
electrophilic reactants or they are converted
metabolically into a chemically reactive electrophilic
form that reacts with some critical nucleophilic site(s) in
macromolecules to initiate the carcinogenic process.
There are a number of important modifying factors in
chemical carcinogenesis such as genetic, aging,
immunological, environmental and hormonal.
Extensive data suggest that chemical carcinogenesis is
a multistage process which can be divided into three
general stages: initiation, promotion, and progression.
The initiation stage appears to be an irreversible step
that probably involves a somatic mutation in some aspect
of growth control and/or differentiation. A good
correlation exists between the carcinogenicity of many
chemical carcinogens and their mutagenic activity.
Several studies have demonstrated a good correlation
between the skin tumor initiating activities of several
tumor initiators and their ability to bind covalently to
DNA. Tumor promotion appears to be reversible for a
relatively long period but later becomes irreversible.
There is quite a diversity of chemical agents that act as
tumor promoters in the various systems where
initiation-promotion has been shown to occur. In terms
of mechanism of action most of the data has been derived
from the use of the phorbol ester tumor promoters in the
mouse skin two-stage tumorigenesis system. The skin
tumor promoters are not mutagenic but bring about a
number of important epigenetic events such as
epidermal hyperplasia, increased dark cells,
membrane changes, altered differentiation, inhibition

0097–6156/89/0414–0078$06.00/0
© 1989 American Chemical Society

of cell-cell communication, embryonic phenotype, and increased prostaglandin and polyamine synthesis. The major effect of the promoters, regardless of the type, is the specific expansion of the initiated stem cells in the skin. This appears to occur by both direct and indirect mechanisms involving the loss of glucocorticoid receptors, differentiation alterations, a direct growth stimulation of the initiated cells and/or selective cytoxicity. The progression stage is characterized by high level of genetic instability which leads to a number of chromosomal alterations. These changes may be responsible for the large differentiation changes such as the loss of the high molecular weight keratin proteins and filagrin increase GGT activity and changes in oncogene expression in squamous cell carcinoma.

Along with epidemiological data, studies using experimental animals have provided evidence that some chemicals in our environment are responsible for a significant proportion of human cancers. The role of diet as an important environmental factor in the etiology of cancer will only be briefly discussed in this chapter and the reader is referred to other reviews (1-2). Shown in Table I are the chemicals that are generally recognized as carcinogens in the human and the sites where they cause tumors. In addition to the chemicals generally recognized as carcinogens in humans as a result of industrial, medical and societal exposures, a number of other chemicals in the environment, such as aflatoxin B, and certain N-nitrosamines and N-nitrosamides are strongly suspected of causing cancer (3-4). It appears likely that additional chemical carcinogens of both natural and synthetic origin will be identified as cancer causing.

Following a discussion of the importance of metabolic activation of chemical carcinogens, I will briefly review a number of general factors that play a modifying role in chemical carcinogenesis. Furthermore, information will be presented suggesting that chemical carcinogenesis is a multistage process which can be divided into three general stages: initiation, promotion, and progression. An important aspect of the multistage theory of cancer induction is that it has suggested that both genetic and epigenetic mechanisms are important in carcinogenesis. Altered growth control and differentiation leading to a more embryonic phenotype are important characteristics of all cancers which may occur by both genetic and epigenetic mechanisms.

<u>Importance of Metabolic Activation of Chemical Carcinogens to Electrophilic Intermediates</u>

A significant general unifying theory to explain the initial event in chemical carcinogenesis has been proposed by the Millers (3,5) which states that all chemical carcinogens that are not electrophilic reactants must be converted metabolically into a chemically reactive electrophilic form that then reacts with some critical nucleophilic site(s) in macromolecules to initiate the carcinogenic process (3,5). Although there have been extensive studies on the metabolic activation of many types of chemical carcinogens, emphasis in this review will be placed on the metabolic activation of polycyclic aromatic hydrocarbons, aromatic amines, and nitrosamines which appear to be highly relevant to human cancer.

Table I. Chemicals Considered as Carcinogens in the Human[a]

Chemical	Sites of Tumor Formation
Industrial exposures	
2(orB)-naphthylamine	Urinary bladder
benzidine(4,4' -diaminobiphenyl)	Urinary bladder
4-aminobiphenyl and 4-nitrobiphenyl	Urinary bladder
bis(chloromethyl) ether	Lungs
bis(2-chloroethyl) sulfide	Respiratory tract
vinyl chloride	Liver mesenchyme
certain soots, tars, oils	Skin, lungs
chromium compounds	Lungs
nickel compounds	Lungs, nasal sinuses
asbestos	Pleura, peritoneum
asbestos plus cigarette smoking	Lungs, pleura, peritoneum
benzene	Bone marrow
mustard gas	Respiratory system
Medical exposures	
N,N-bis(2-chloroethyl)-2-naphthylamine	
(chlornapthazine)	Urinary bladder
diethylstilbestrol	Vagina
estrogenic steroids	Breast and uterus
Societal	
cigarette smoke	Lungs, urinary tract, pancreas
betel nut and tobacco quids	Buccal mucosa

[a]Reviewed in reference 4

The polycyclic aromatic hydrocarbon carcinogens, which are very ubiquitous, are metabolized by the microsomal mixed-function oxidase system of target tissues to a variety of metabolites such as phenols, quinones, epoxides, dihydrodiols and dihydrodiol-epoxides (6). The major pathway of activation of benzo(a)pyrene (BP) leads to the formation of dihydrodiol-epoxide of BP which interacts predominantly with the 2-amino of guanine of DNA. The dihydrodiol-epoxide of BP appears to be the major ultimate electrophilic, mutagenic, and carcinogenic metabolite of BP (6). Nevertheless, other metabolites such as certain phenols, epoxides and quinones may contribute to the overall carcinogenic activity of BP. In addition, a free radical mechanism may also be partly involved in its carcinogenic activity.

Aromatic amines comprise a large family of compounds, both industrial and medicinal as well as consumer products. They possess reactive primary or secondary amine groups that play an active role in determining the carcinogenicity of the compounds (5). Examples of aromatic amine carcinogens are aniline, 2-naphthylamine, benzidine, p-dimethylaminoazobenzene (butter yellow), and N-2-fluorenylacetamide (2-FAA). The metabolic activation of 2-FAA involves a N-hydroxylation (proximate carcinogen) followed by esterification to an ultimate reactive carcinogen that can react with nucleic acids or proteins. In terms of DNA, the major adduct occurs at the C-8 of guanine (5).

The nitrosamines and nitrosamides are an important class of chemicals since a large number have been found to be carcinogenic. Nitrosamines require metabolic activation by target tissue microsomal mixed-function oxidases. Unlike polycyclic aromatic hydrocarbons, nitrosamines require only the first activation step which forms a hydroxylated intermediate that is sufficiently unstable to spontaneously decompose and generate a reactive carbonium which can rapidly alkylate target macromolecules in the cell. In contrast, nitrosamides do not require metabolic activation because of inherent chemical instability in aqueous solution. Nitrosamides decompose at physiological pH nonenzymatically to produce the same class of reactive electrophiles as the nitrosamines. Although the actual target site for nitroso carcinogens is not yet elucidated, a number of alkylation sites in DNA are known such as O^6-CH$_3$-G, 7-CH$_3$-G, and 3-CH$_3$-A (3).

Since the synthesis of N-nitroso compounds is accomplished through reaction of secondary amines and nitrous acid, it was demonstrated that nitrosamines could be produced biochemically by the naturally acidic conditions in the stomach with ingestion of secondary amines (3). The nitrosating agent would be formed by the reaction of gastric juice with nitrite compounds that are widely used as food preservatives and for food color enhancement (3). Vitamin C was found to be a potent inhibitor of this reaction (7).

<u>Modifying Factors of Chemical Carcinogenesis</u>

A number of factors have been shown to play a modifying role in chemical carcinogenesis in experimental animals, and more than likely, play similar roles in human cancer (Table II). Chemical carcinogens can have either an additive, synergistic, or an inhibitory effect on the carcinogenic activity of a given carcinogen (4). There are agents that can inhibit chemical carcinogenesis by counteracting the metabolism of the carcinogen as well as by increasing the metabolism of the carcinogen but favoring detoxification. This will be discussed in more detail later in this chapter when some anticarcinogenic agents are discussed.

Table II. General Factors or Modifiers of Chemical Carcinogenesis

1.	Other chemical carcinogens
2.	Metabolism of carcinogen (activation vs detoxification)
3.	Anti-carcinogenic chemicals
4.	DNA repair
5.	Age
6.	Sex
7.	Hormonal
8.	Immunological
9.	Trauma
10.	Radiation
11.	Viral
12.	Cocarcinogenesis and tumor promotion
13.	Diet, nutrition and life style
14.	Genetic constitution

An important determinant of whether or not a cell becomes initiated by a chemical carcinogen is DNA repair. Inhibition of the excision repair system allows a greater chance that the carcinogenic damage will not be repaired, thus leading to the irreversible initiated carcinogenic state (4). Likewise, inhibition or induction of error prone DNA repair could lead to a drastic modification of the carcinogenesis process (4).

An important modifying influence on carcinogenesis is age with the noted correlations between the incidence of cancer and aging of a population (8). Van Duuren and his associates demonstrated a small but general decrease in the rate of tumor production with an increasing age at the time of tumor promotion (9). On the other hand, when the mouse skin of both young and old animals was treated with several applications of 7,12-dimethylbenz(a)anthracene (DMBA) and then grafted to young recipients, a higher incidence of carcinoma developed in grafts from old donors than from young (10). These experiments suggest that adult tissue is more prone toward tumor initiation because of either a faulty DNA repair system or a greater DNA error prone system and is less susceptible to tumor promotion whereas the reverse appears to be true in newborn tissue.

Tumor growth can be modified by sex hormones as well as hormones in general. Mammary tumors in rats are essentially always hormone dependent. For example, breast tumors in rats regress following the removal of the ovaries showing the cocarcinogenic effect of female sex hormones (11-12). A similar situation has been observed with prostate cancer and the male sex hormones (11). Besides the sex hormones, the pituitary is decisive in tumor growth. Instead of removing the ovaries, one can extirpate the pituitary in order to stop mammary carcinomas. It is also well known that the sex of the animal is important in carcinogenesis by certain chemicals. 2-FAA is a potent liver carcinogen in male rats but not females (3). If the testes and thyroid are removed from rats, they will not develop liver cancer; however, 2-FAA will induce hepatomas after supplementary dosing with testosterone and thyroxine (12).

The immunological status of the animal is also an important factor in chemical carcinogenesis. It has been shown that Baccillus Calmette-Guerin (BCG) vaccination reduces the carcinogenic effect of both carcinogens and promoters (13-

14). BCG vaccination may cause its effect on neoplasms through the nonspecific enhancement of the immunologic surveillance mechanism of the host (14). The establishment and proliferation of malignant neoplasms may result from the inability of the host to recognize the tumor immunologically as foreign and thus destroy it (15). Experiments have shown that substances that alone cause cancer and substances that can act only to promote cancer, such as croton oil and phorbol ester, are immunosuppressive agents (16-18). The potent antipromoting steroids are also potent immunosuppressive agents (19). However, the antipromoting ability of glucocorticoids appears to be related to some early effect on tumor promotion not involving the immune system. In fact, if the glucocorticoids are applied topically to existing carcinomas and papillomas, they appear to enhance their progression (19). The demonstration of tumor associated transplantation antigens in the majority of the malignancies studied (15,18) and the presence in the tumor-bearing host of concomitant immune reactions against those antigens (15) have led to theories suggesting that the immune system plays a major role as a protection against malignancy (18). Thus, in all the presentations of the "immune surveillance" theory, the effects of immunodepression (thymectomy and other procedures and drugs) favoring tumor development are quoted as experimental evidence that some sort of control of tumor development is exerted by the intact immune system (18). Such immune surveillance theories would predict a high risk of tumor development in the athymic-nude mice. However, Stutman's results do not support that prediction, since tumor incidence after exposure to 3-methylcholanthrene (MCA) at birth was similar in the immunologically normal nude hyterozygotes and the immunologically deficient athymic-nude homozygotes (18).

Trauma appears to play an important role in cancer. Any long-lasting wound or sore is considered a potential site for cancer and is thus listed as one of the warning signs. It has been shown that wound healing alone can promote the formation of skin tumors initiated by DMBA (20). As pointed out later in this chapter, inflammation and cellular proliferation are related to chemically induced cancer. Chemical carcinogens and tumor promoters induce inflammation and hyperplasia in mouse skin, but it is also known that not all inflammatory-hyperplastic agents cause skin cancer or promote it (21).

Radiation alone or in combination with carcinogenic chemicals has been known for a long time to be involved in the etiology of skin cancer. A correlation has been noted between the occurrence of skin cancer and ultraviolet light exposure with the incidence notably higher in sunny regions, among outdoor workers, and among people with fair complexions, especially on more exposed skin areas (22). The first report of a suspected interaction between radiation and chemical carcinogens in humans was related to the high incidence of skin cancer among sailors which was thought to be caused by a combination of the high doses of sunlight and coal tar to which they were exposed. Epidemiological studies of uranium miners have indicated an interaction between ionizing radiation and cigarette smoke (23). Both alone and to some extent in combination with chemical carcinogens and tumor promoters, the role of both ultraviolet and ionizing radiation in the induction of skin cancer, as well as other cancers in experimental animals, has been extensively studied (22,24-25). It is quite apparent from these studies that when either ultraviolet or ionizing radiation is combined with chemical carcinogens, the times of tumor appearance are accelerated and/or the tumor incidence are increased. Although both ultraviolet and ionizing radiation have been shown to be either promoting or enhancing agents for chemically initiated tumors and initiating agents for chemically promoted tumors, their primary influence appears to be related to their tumor-promoting activities rather than initiation.

Studies on the interaction between viruses and chemical carcinogens were carried out soon after the first tumor viruses were isolated (26). Such interactions generally result in a higher incidence of tumors in animals exposed to both agents as compared to animals exposed to either agent alone. Although studies of the induction of cancer by a combination of chemicals, viruses, and radiation have not been undertaken, such investigations may indeed reveal a closer relationship among chemical, viral, and radiation carcinogenesis.

Important modifying factors of chemical carcinogenesis are diet and nutrition in experimental animals as well as in cancer in humans (2). Diet appears to be more a cocarcinogen and/or a tumor promoter in carcinogenesis rather than a causative agent. Epidemiological studies have revealed that there are notable differences in the incidence of specific cancers from country to country and even within countries (1,27). These differences in the geographic incidence of cancer do not appear to be primarily genetically determined since the cancer patterns for migrants from one country to another generally change from those characteristics of the mother country to those of the inhabitants of the new country.

Although the above studies suggest that the genetic makeup of an individual appears not to be of primary importance in certain cancer patterns, it is well known that the genetic constitution of a person can make him prone to cancer. Not to be mistaken as a contradiction, the changing patterns of cancer related to migrants appear to be related to the diet and life style of the individual which act more as cocarcinogens and/or tumor promoters (28). If a migrant adopts the cancer-related diet of a country and is genetically prone to cancer, there probably will be a good chance that the migrant will develop the diet-related cancer; however, if the migrant was not genetically prone, he probably would not develop the diet-related cancer. In the above situation the new diet promoted the cancer. Since we are all more than likely initiated because of the large number of initiating agents present in our environment, promotion is probably the rate limiting factor in cancer.

There are several genetically determined diseases such as xeroderma pigmentosum (XP), Fanconi's anemia, Bloom's syndrome, ataxiatelangiectasia and porokeratosis Mibelli in which individuals have an increased if not an invariable incidence of cancer (29). These diseases are associated with DNA repair defects, thought to be the cause of the cancer sensitivity (29). In the case of a person with XP, the oversensitivity of the skin to ultraviolet light is inherited and the condition leads to skin cancer.

Mice can be bred with high and low sensitivity to chemical carcinogens. Boutwell (11) accomplished this by stressing mice with a DMBA-promoted two-stage system of tumorigenesis and then selecting for sensitivity and resistance to the carcinogenic treatment for eight generations. The outcome was a remarkable sensitivity and resistance to the chemical induction of cancer. These and other experiments lead us to conclude that the chemical induction of cancer is also dependent on the genetic constitution of the animal (29). As with spontaneous tumors, genetic constitution can mean many things, among them the loss of the capability to activate carcinogens, faulty DNA repair, an incompetent immune system, and an increased sensitivity to growth factors leading to a greater turnover of cells in a given tissue.

Multistage Carcinogenesis

It is well known that chemical carcinogenesis is a two-stage or multistage process with one of the best studied models being the two-stage carcinogenesis system using mouse skin (4). Skin tumors can be induced by the sequential application of a subthreshold dose of a carcinogen (initiation stage) followed by repetitive treatment with a noncarcinogenic promoter (promotion stage). The initiation phase requires only a single application of either a direct or an indirect carcinogen and is essentially an irreversible step, while the promotion phase is initially reversible later becoming irreversible. A single large dose of a carcinogen such as DMBA is capable of inducing skin tumors in mice. Papillomas occur after a relatively short latency period (10 to 20 weeks) with carcinomas developing after a much longer period (20 to 60 weeks). If this dose is lowered, it becomes necessary to repeatedly administer DMBA in order to induce tumors. If progressively reduced, a subthreshold dose of DMBA is reached which will not give rise to tumors over the lifespan of the mouse. If either croton oil or a phorbol ester such as 12-0-tetradecanoylphorbol-13-acetate (TPA) is subsequently applied repetitively to the backs of mice previously initiated with a single subthreshold dose of DMBA, multiple papillomas will appear after a short latency period followed by the appearance of squamous cell carcinomas after a much longer period. The repetitive application of the promoter, without initiation by DMBA, in general either does not give rise to tumors or produces only a few but never exhibiting a dose-response relationship (30). If the mice are initiated with a subthreshold dose of a carcinogen such as DMA, there is an excellent dose-response using TPA as the promoter (30). Likewise, there is a very good dose-response with BP or DMBA as a tumor initiator when the promoter dose is held constant (31). If repetitive applications of the promoter are administered before initiation, no tumors will develop. The real hallmark of the two-stage carcinogenesis system in mouse skin relates to the irreversibility of tumor initiation. Even a lapse of up to one year between the application of the initiator and the beginning of the promoter treatment provides a tumor response similar to that observed when the promoter is given only one week following initiation (31). Unlike the initiation phase, the promotion stage is reversible and requires a certain frequency of application in order to induce tumors (31).

Tumor Initiation. The skin tumor initiation stage appears to be an irreversible step that probably involves a somatic mutation in some aspect of epidermal growth control and/or epidermal differentiation (31). Extensive data have revealed a good correlation between the carcinogenicity of many chemical carcinogens and their mutagenic activities (31). Most tumor-initiating agents either generate or are metabolically converted to electrophilic reactants, which bind covalently to cellular DNA and other macromolecules (33, 31). Previous studies have demonstrated a good correlation between the skin tumor initiating activities of several polycyclic aromatic hydrocarbons and their abilities to bind covalently to DNA (31). Slaga and Klein-Szanto have recently found that skin tumor initiation probably occurs in dark basal keratinocytes since a good correlation exists between the degree of tumor initiation and the number of dark basal keratinocytes present in the skin (32). The dark basal keratinocytes are present in the skin in large numbers during embryogenesis, in moderate numbers in newborns, in low numbers in young adults, and in very low numbers in old adults suggesting these cells may be epidermal stem cells (33). The initiating potential of mouse skin decreases with the age of the mouse to the point that it is very difficult to initiate mice greater than one year of age when the number of dark basal keratinocytes are extremely rare. Evidence also suggests that skin tumor initiators interact in a specific way with slowly cycling, self-renewing stem cells in the epidermis (34). These cells are also a very dense keratinocyte but the relationship with epidermal dark cells is presently not known.

Tumor Promotion. The extensiveness of the data available as well as the multistage nature of tumor promotion suggests that this process, which is now thought to occur in most tissues where cancer can be induced or where it occurs spontaneously, may involve the interaction of a number of endogenous factors as well as environmental factors such as chemicals, radiation, viruses, and diet and nutrition, thus unifying all current areas of cancer research (4). In human cancer, smoking, asbestos, radiation, hormones, and diet and nutrition, to mention a few, are now thought to have more of a promotional influence in the multistage carcinogenesis process (4).

A variety of chemical agents are known to act as tumor promoters in various systems. However, much of our knowledge of the cellular and molecular mechanisms of tumor promotion has come from studies using the phorbol ester tumor promoters in the mouse skin model and in various cell culture systems.

Current information suggests that skin tumor promoters do not bind covalently to DNA and are not mutagenic but bring about a number of important epigenetic changes (30). Of the observed phorbol ester-related effects on the skin, the induction of epidermal cell proliferation, ornithine decarboxylase (ODC) and subsequent polyamines, prostaglandins and dark basal keratinocytes have the best correlation with promoting activity (31). In addition to the induction of dark cells, which are normally present in large numbers in embryonic skin, many other embryonic conditions appear in adult skin after treatment with tumor promoters. Most of the embryonic conditions which occur after tumor promoter treatment may be a consequence of the alteration in differentiation.

Skin tumor promoters bring about a number of other important epigenetic changes in the skin such as membrane and differentiation alterations and an increase in protease activity, cAMP independent protein kinase activity and phospholipid synthesis (31). In addition, the skin tumor promoters cause a decrease in epidermal superoxide dismutase and catalase activities as well as a decrease in the number of glucocorticoid receptors (31). Some skin promoters appear to have a common mode of cellular action - via binding to the natural cellular substrate for diacylglycerol-a phospholipid, calcium-dependent kinase called protein kinase C. Promoters which interact with protein kinase C include 12-0-tetradecanoylphorbol-13-acetate and related phorbol agents, teleocidin and its analogs and aplysiatoxins (31). Other promoters such as benzoyl peroxide, lauroyl peroxide, hydrogen peroxide, anthralin, chrysarobin, and palytoxin do not interact with the protein kinase C and apparently act as promoters via some other mechanism, possibly their ability to generate free radicals. Even with the ability to classify promoters based on their interaction with protein kinase C, the mechanisms involved in promotion are not clearly understood.

Ultimately, the major effect of all promoters, regardless of type, is the expansion of the initiated cell to form visible tumors. Table III lists possible mechanisms for the selection of initiated epidermal cells by tumor promoters. Of the four mechanisms mentioned, evidence supporting each can be found experimentally. Recent studies by Morris (34) have demonstrated that slowly cycling basal cells (thymidine label retaining cells) are induced to divide by a single dose of TPA. Many skin tumor promoters increase the presence of dark basal keratinocytes in the epidermis which correlates well with the activity of first stage tumor promoters such as TPA, hydrogen peroxide, calcium, ionophore A23187 (33). The significance of dark basal keratinocytes is not clear, but they appear normally in large numbers in embryonic skin, and their presence in promoted skin may reflect environmental changes within the epidermis which allow the expansion of cells (including initiated cells) similar to that observed in embryonic tissue (33). Tumor promoters are potent inducers of terminal differentiation (35) and mechanistically may cause the expansion of initiated cells via a decrease in negative feedback signals on cell proliferation. Tumor promoters, especially the free-radical generating promoters such as benzoyl peroxide and hydrogen peroxide, are highly cytotoxic and may promote by causing a regenerative hyperplasia in the skin similar to the response observed after wounding.

Table III. Mechanisms of Selection of Initiated Epidermal Stem Cells by Skin Tumor Promoters

---

1. Some tumor promoters may have direct effect on initiated stem cells (dark cells?) causing them to divide and expand in number.
2. Tumor promoters convert some basal keratinocytes to an embryonic phenotype similar to the dark cells thereby supplying a positive environment for the dark cells to expand in number.
3. Tumor promoters stimulate terminal differentiation of some epidermal cells and thus decrease a negative feedback mechanism on cell proliferation.
4. Some tumor promoters have a selective cytotoxic effect which may cause initiated cells to expand in number.

Tumor Progression. Although many studies have been directed toward understanding the mechanisms involved in skin tumor initiation and promotion, orly a few studies have been performed on the progression stage of skin carcinogenesis. Table IV summarizes some of the important characteristics of skin papillomas and carcinomas in order to emphasize events which appear to be critical in the conversion of papillomas to carcinomas and other aspects of tumor progression. As can be seen, there are several events which appear during tumor promotion that are continued or even exaggerated during tumor progression such as an increase in dark cells, loss of glucocorticoid receptors and an increase in polyamines and prostaglandins (36).

A number of changes occur very late in the carcinogenesis process which are related to the conversion of benign to malignant tumors. We have found that all squamous cell carcinomas lack several differentiation product proteins such as high molecular weight keratins (60,000 - 62,000) and filaggrin but are positive for gamma glutamyltransferase (GGT), whereas only about 20% of the papillomas generated by an initiation-promotion protocol exhibit a similar condition (37-38). Before visible tumors are observed using the initiation-promotion protocol, these conditions appear normal suggesting that they are very late responses (38).

Balmain and coworkers (39) have recently found that a percentage of papillomas and carcinomas induced by DMBA-TPA contained elevated levels of Ha-ras transcripts compared with normal epidermis. Furthermore, the tumor

Table IV. Characteristics of Skin Tumors

Benign Papillomas

1.   Large number of dark cells.
2.   Loss of glucocorticoid receptors.
3.   High level of polyamines and prostaglandins.
4.   Approximately 80% of the papillomas induced by two stage protocol have
     high molecular weight keratins and filaggrin and are negative for GGT.
     20% have reverse conditions.
5.   Approximately 50% of papillomas induced by the two stage protocol express
     Ha-ras RNA.
6.   Some papillomas are reversible while others are irreversible.
7.   Early papillomas (10 wks) during promotion are well differentiated
     hyperplastic lesions with either mild or no cellular atypia, whereas late
     ones (40 wks) are dysplastic, show atypia and are aneuploid.
8.   Sequential appearance of trisomy of chromosome 6 followed by trisomy of
     chromosome 7.

Carcinomas

1.   Large number of dark cells.
2.   All lack glucocorticoid receptors.
3.   High level of polyamines and prostaglandins.
4.   All lack high molecular weight keratins and filaggrin.
5.   All are positive for GGT.
6.   Approximately 67% of carcinomas induced by two-stage protocol express
     Ha-ras RNA. Complete carcinogenesis protocol by MNNG does not give
     increase in expression of Ha-ras RNA but increase expression of src and
     abl.
7.   All are aneuploid with some non-random chromosomal changes such as
     trisomy in chromosome 6, 7, and 2.

DNA was capable of malignantly transforming NIH 3T3 cells in DNA transfection studies (40). Studies in our laboratory (41) indicate that initiation alone or repetitive TPA treatments are insufficient to turn on the expression of the Ha-ras oncogene in adult SENCAR mouse epidermis. Initiation followed by either one or six weeks of TPA treatment also failed to activate Ha-ras expression. Like Balmain, we observed elevated levels of Ha-ras RNA in a percentage of papillomas and carcinomas tested. We have also found that the expression of c-src and c-abl are increased in the majority of carcinomas examined (42). It remains to be determined whether oncogene activation plays a critical role in multistage skin carcinogenesis.

Hennings and coworkers (43) recently reported that if mice with papillomas are treated repetitively with N-methyl-N-nitro-n-nitrosoguanidine (MNNG), a significant increase in the conversion of papillomas to carcinomas occurs. We have also found similar results with limited treatment with MNNG as well as with ethylnitrosourea (ENU), benzoyl peroxide and hydrogen peroxide (44-45). This type of treatment (initiation-promotion-initiation) produces a carcinoma response similar to complete carcinogenesis, i.e. the repetitive application of a carcinogen such as DMBA or MNNG probably continuously supplies both initiating and promoting influences. An explanation of why a different type of promoter, such as benzoyl peroxide, or a non-promoter, such as hydrogen peroxide, can increase the conversion of papillomas to carcinomas is presently not understood (46).

The mechanisms involved in progression in the mouse skin system are unclear. The carcinogens, ENU and MNNG, and the peroxides are all genotoxic compounds. Chromosomal studies have shown that squamous cell carcinomas are highly aneuploid lesions often exhibiting hyperdiploid stem cell lines (47). Although early papillomas (10 wks of promotion) are diploid, they progressively show chromosomal changes and eventually all become aneuploid after 30 to 40 weeks of promotion (47). Additional evidence does indicate specific chromosome alterations. Using a direct cytogenetic technique, we identified a non-random trisomy of chromosome 6 in 100% of aneuploid mouse skin papillomas and in 10 of 11 squamous cell carcinomas induced by chemical carcinogenesis. The second most common abnormality observed was trisomy of chromosome 7 found in most dysplastic papillomas and 9 of 11 carcinomas (48). Trisomy of chromosome 6 occurred before trisomy of chromosome 7 (48). Both trisomies were the only abnormalities found in all aneuploid papillomas and in several carcinomas (48). More progressed carcinomas also had trisomies of chromosome 2 and 13. Whether the genotoxic effects of the agents used in progression experiments are able to induce such specific alterations are presently unknown.

In addition to chromosomal alterations, squamous cell carcinomas exhibit changes in protein expression, including the lack of high molecular weight keratins (49), filaggrin (50) and the presence of GGT. Possibly these phenotypic changes are the result of the gene alterations and rearrangements and can be induced by genotoxic agents. Histological and cytochemical studies of keratoacanthomas induced by both ENU and benzoyl peroxide, when these agents were used as "progressors", showed a high percentage of GGT-positive tumors possibly reflecting a novel expression of this enzyme in benign lesions (30, 51).

A different mechanism of genetic alteration which could be relevent to progression is change in the methylation state of DNA. Preliminary evidence from this laboratory has indicated that a gradient in DNA methylation exists from normal mouse epidermis to papillomas to carcinomas with carcinomas being highly under-methylated (La Peyre et al., unpublished results). Concerning the

agents used as "progressors" in these studies, both ENU and MNNG have been
shown to inhibit methylation by blocking DNA methyltransferase activity (40, 52).
An alternate hypothesis for the action of "progressor" agents relates to their
high degree of cytotoxicity. In this model for progression, highly cytotoxic agents
may 1) selectively or nonselectively kill cells within a tumor allowing the growth
of more malignant cells and/or 2) kill normal cells reducing the constraints
against expansion along the border between normal and tumor tissue. Both
alternatives assume that cells capable of invasion either pre-exist within the
benign tumor or that expansion of tumor clones increases the chance of the natural
progression of cells toward malignancy. We are currently testing a number of
cytotoxic agents for activity during the progression stage.

Inhibitors of Carcinogenesis

Although prevention of the exposure of man to carcinogens is theoretically the best
way to reduce cancer incidence, such an approach is not practical for obvious
reasons. Therefore, alternate means of modifying the process of carcinogenesis in
man must be found and in view of the fact that carcinogenesis is a prolonged
multistage process, a variety of approaches may be considered toward the inhibition
of either the initiation or the promotion phases. Table V summarizes various
general classes of chemicals used to inhibit chemical carcinogenesis, tumor
initiation, tumor promotion and/or tumor progression (31).

Table V. General Classes of Chemicals that Inhibit Chemical Carcinogenesis

Antioxidants and free radical scavenging agents
Vitamins (A, C, E)
Protease inhibitors
Retinoids
Flavones
Anti-inflammatory steroids
Certain noncarcinogenic PAH and environmental contaminants
Polyamine synthesis inhibitors (DMFO, retinoids)
Prostaglandin synthesis inhibitors
Chemicals that alter cyclic nucleotide levels
Glutathione

        Using the skin tumorigenesis system, one can specifically study the effects
of potential inhibitors on the initiation, promotion, and progression phases (31).
Studies have been performed on many compounds that have the capacity to inhibit
tumor initiation by either 1) alteration of the metabolism of the carcinogen
(decreased activation and/or increased detoxification), 2) scavenging of active
molecular species of carcinogens to prevent their reaching critical target sites in
the cells, or 3) competitive inhibition. In addition, there have been a number of
studies of compounds that inhibit the promotion or the progression of cancer either
by altering the state of differentiation, by inhibiting the promoter-induced cellular
proliferation or by preventing gene activation by the tumor promoters. Presently,
only glutathione has been shown to effectively inhibit the tumor progression stage,

suggesting the involvement of free radicals. Figure 1 illustrates the potential sites of action of inhibitors of chemical carcinogenesis.

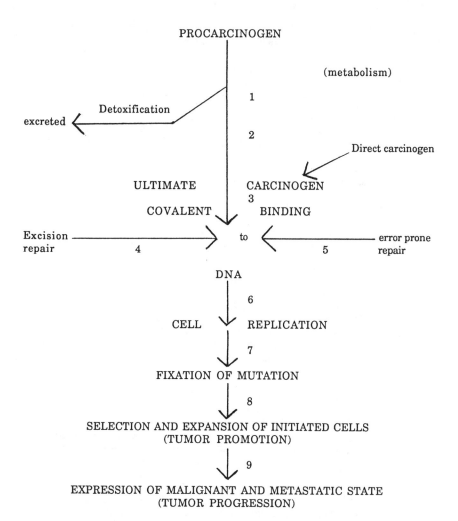

Figure 1:   Illustration of potential sites of action of inhibitors of chemical carcinogenesis. Numbers 1 - 9 represent sites where inhibitors may be effective.

Acknowledgments

Research was supported by the Public Health Service Grant CA43278 from the National Cancer Institute and the Olga Keith Weiss Chair. I gratefully acknowledge Karen Engel, Christie Hoy and Mary Slaga for their assistance.

Literature Cited

1.   Armstrong, B., and Doll, R. Int. J. Cancer. 1975, 15, 617-631.
2.   Wynder, E.L. Fed. Proc. 1976, 35, 1309-1315.
3.   Miller, E.C. Cancer Res. 1978, 38, 1479-1496.
4.   Slaga, T.J. In Modifiers of Chemical Carcinogenesis;Slaga, T.J., Ed.;
     Raven Press, New York, 1980; pp. 243-262.
5.   Miller, E.C., and Miller, J.A. In Chemical Carcinogens;Searle, C.E.,
     Ed.; American Chemical Society, Washington, DC, 1976; p. 732.
6.   Jerina, D.M., Lehr, R.D., Yagi, H., Hernandez, O., Dansette, P.M.,
     Wislocki, P.G., Wood, A.W., Chang, R.L., Levin, W. and Conney, A.H.
     In In Vitro Metabolic Activation and Mutagenesis Testing;deSerres, F.J.,
     Fouts, J.R., Bend, J.R., and Philpot, R.M., Ed.; Elsevier/North Holland
     Biomedical Press, Amsterdam, 1976; pp. 159-177.
7.   Mervish, S.S., Wallcave, L., Eagen, M., and Shubik, P. Science. 1972, 177,
     65-68.
8.   Pitot, H.C. Fed. Proc. 1978, 37, 2841-2847.
9.   Van Duuren, B.L., Sivak, A., Katy, C., Seidman, I., and Melchionne, S.
     Cancer Res. 1975, 35, pp.502-505.
10. Ebbesen, P. Science. 1974, 183, 217-220.
11. Boutwell, R.K. Prog. Exp. Tumor Res. 1964, 4, 207-250.
12. Heidelberger, C. Chemical Carcinogenesis. 1975, 44, 79-121.
13. Piessens, W.F., Heimann, R., Legros, N., and Heuson, J.C. Cancer Res.
     1971, 31, 1061-1065.
14. Schinitsky, M.R., Hyman, L.R., Blazkovec, A.A., and Burkholder, P.M.
     Cancer Res. 1973, 33, 659-663.
15. Keast, D. Immunosurveillance and Cancer. 1970, 2, 710-712.
16. Bluestein, H.G., and Green, I. Nature. 1970, 228, 871-872.
17. Stjernsward, J. J. Natl. Cancer Inst. 1967, 38, 515-526.
18. Stutman, O. Science. 1974, 183, 534-536.
19. Schwarz, J.A., Viaje, A., Slaga, T.J., Yuspa, S.H., Hennings, H., and
     Lichiti, U. Chem. Biol. Interact. 1977, 17, 331-347.
20. Hennings, H., and Boutwell, R.D. Cancer Res. 1970, 30, 312-320.
21. Slaga, T.J., Fischer, S.M., Viaje, A., Berry, D.L., Bracken, W.M.,
     LeClerc, S., and Miller, D.L. In Mechanisms of Tumor Promotion and
     Cocarcinogenesis; Slaga, T.J., Sivak, A., and Boutwell, R.K., Ed.;Raven
     Press, New York, 1978; pp. 173-195.
22. Emmett, E.A. CRC Critical Rev. Toxicol. 1973, 2, 211-255.
23. Bair, W.J. In Inhalation Carcinogenesis; Nettesheim, P., Hanna, M.G.,
     Jr., and Gilbert, J.R., Ed.; USAEC, Division of Technical Information,
     Symposium Series 18, National Technical Information Service, 1970,pp.77-
     97.
24. Storer, J.B. In Cancer; Becker, F.F., Ed.;Plenum Press, New York, 1975;
     Vol. 1.
25. Upton, A.C. In Methods in Cancer Research; Busch, H., Ed.; Academic
     Press, New York, 1968, Vol 4.
26. Casto, B.C., and DiPaolo, J.A. Prog. Med. Virol. 1973, 16, 1-47.
27. Doll, R. Cancer. 1969, 23, 1-8.

28.   Wynder, E.L., Hoffmann, D., McCoy, G.D., Cohen, L.A., and Reddy, B.S. In Mechanisms of Tumor Promotion and Cocarcinogenesis; Slaga, T.J., Sivak, A., and Boutwell, R.K., Ed.; Raven Press, New York, 1978; pp. 59-77.

29.   Frei, J.V. Chem. Biol. Interact. 1976, 13, 1-25.

30.   Slaga, T.J. Overview of Tumor Promotion in Animals Environmental Health Perspectives. 1983, 50, 3-14.

31.   Slaga, T.J., Fischer, S.M., Weeks, C.E., Klein-Szanto, A.J.P., Reiners, J. J. Cellular Biochem. 1982, 18, 99-119.

32.   Slaga, T.J., and Klein-Szanto, A.J.P. Cancer Investigation. 1983, 1(5), 425-436.

33.   Klein-Szanto, A.J.P., Slaga, T.J. Cancer Res. 1981, 41, 4437-4440.

34.   Morris, R.J., Fischer, S.M., and Slaga, T.J. Cancer Res. 1986, 46, 3061-3066.

35.   Reiners, J.J., Slaga, T.J. Cell. 1983, 32, 247-255.

36.   Slaga, T.J. Cancer Surveys. 1983, 2(4), 595-612.

37.   Klein-Szanto, A.J.P., Nelson, R.G., Shah, Y., and Slaga, T.J. J. Natl. Cancer Inst. 1983, 70, 161-168.

38.   Nelson, K.G., Stephenson, K.B., Slaga, T.J. Cancer Res. 1982, 42, 4164-4795.

39.   Balmain, A., Ramsden M., Bowden, G.T., and Smith, J. Nature. 1984, 307, 658-660.

40.   Balmain, A., and Pragnell, I.D. Nature. 1983, 303, 72-74.

41.   Pelling, J.C., Hixson, D.C., Nairn, R.S., and Slaga, T.J. Proc. Amer. Assoc. Cancer Res. 1984, 25, p. 78.

42.   Patskan, G.J., Pelling, J.C., Nairn, R.S. and Slaga, T.J. Pennsylvania State University Fourth Summer Symposium in Molecular Biology, 1985.

43.   Hennings, H., Shores, R., Wenk, M.L., Spangler, E.F., Tarone, R., and Yuspa, S.H. Nature. 1983, 304, 67-69.

44.   O'Connell, J.F., Klein-Szanto, A.J.P., DiGiovanni, D.M., Fries, J.W., and Slaga, T.J. Cancer Res. 1986, 46, 2863.

45.   Parry, J.M., Parry, E.M., and Barrett, J.C. Nature. 1981, 294, 263-265.

46.   Rotstein, J.B., O'Connell, J.F., and Slaga, T.J. Proc. Am. Assoc. Cancer Res., 1986, 27, 143.

47.   Aldaz, C.M., Conti, C.J., Klein-Szanto, A.J.P., and Slaga, T.J. Proc. Natl. Acad. Sci. U.S.A., in press.

48.   Aldaz, C.M., Trono, D., Larcher, F., Slaga, T.J., and Conti, C.J. Molec. Carcin. 1989, 2, 22-26.

49.   Klein-Szanto, A.J.P., Nelson, R.G., Shah, Y., and Slaga, T.J. J. Natl. Cancer. Inst. 1983, 70, 161-168.

50.   Mamrack, M.D., Klein-Szanto, A.J.P., Reiners, Jr., J.J. and Slaga, T.J. Cancer Res. 1984, 44, 2634.

51.   O'Connell, J.F., Klein-Szanto, A.J.P., DiGiovanni, D.M., Fries, J.W., and Slaga, T.J. Cancer Res. 1986, 30, 269.

52.   Wilson, V.L. and Jones, P.A. Cell. 1983, 32, 239.

RECEIVED July 28, 1989

Chapter 6

Pesticide-Induced Modulation of the Immune System

Peter T. Thomas and Robert V. House

IIT Research Institute, 10 West 35th Street, Chicago, IL 60616–3799

The immune system is a recognized target organ for the
toxicologic effects of pesticides. Studies in animals
have documented immune dysfunction following relatively
short-term exposure to pesticides leading to an
increased susceptibility to infection and, arguably,
cancer. However, other than hypersensitivity
reactions, the evidence in humans linking exposure to
pesticides and adverse health effects associated with
immune dysfunction is inconclusive at this time.

Knowledge of the immune system's role in the maintenance of health
has increased dramatically over the last decade. As a result, appreciation of the system as an important and sensitive target organ for
toxicity has also grown. The immune system is composed of a complex
set of cellular and soluble components that protect the individual
against foreign ("nonself") agents while not responding adversely to
"self" tissues. The distinction between self and nonself is made by
an elaborate recognition system that depends on specific receptor
molecules associated with certain immune cells including T- and B-
lymphocytes. Optimal functioning of the immune system requires that
these cells, cell products, and regulatory proteins interact with
each other in a sequential, regulated manner (Figure 1). Other cell
types and nonspecific mechanisms that interact with and regulate T-
and B-lymphocyte functions are also important in the immune response.
These cell types and nonspecific systems include mononuclear phago-
cytes, natural killer cells, polymorphonuclear leukocytes, and the
complement system.

The immune system plays a major role in protecting the host from
infectious disease and, arguably, from cancer. This is demonstrated
by the association between the therapeutic use of chemical immunosuppressants (i.e., in cancer chemotherapy or organ transplantation)
and an increased incidence of infections (1) and certain cancers (2).
This relationship is also illustrated by the Acquired Immune
Deficiency Syndrome (AIDS), in which a loss of immune responsiveness
is associated with infection with Pneumocystis carinii and other

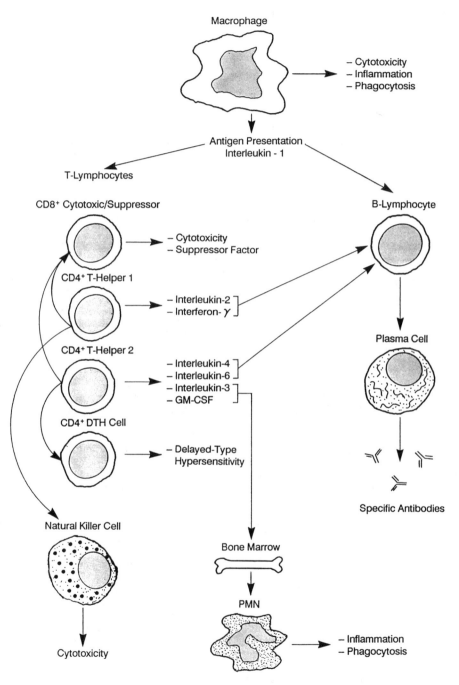

Figure 1. Cellular and humoral interactions in acquired immunity.

opportunistic pathogens, as well as the development of a rare form of cancer known as Kaposi's sarcoma (3).

A wide variety of chemicals and drugs have been shown to affect the immune system adversely, including pesticides (4). Studies with the polycyclic aromatic hydrocarbons (PAHs) have demonstrated that most carcinogenic PAHs are immunosuppressive, whereas their noncarcinogenic congeners are not (5); see review (6). Thus, it follows that exposure to carcinogenic pesticides could potentially result in damage to the immune system.

Within the last 10 years, investigative toxicology has expanded along organ- and system-specific lines. The immune system is now regarded as one of these systems, although until recently its multicomponent nature prevented toxicologists from gaining a true appreciation of it as an integrated target organ for potential toxic damage. In recent years the methodology for assessing immune function has become more standardized and has been validated to the point where routine testing requirements are now being considered by regulatory agencies.

Because pesticides are biologically active chemical compounds, concerns exist regarding their toxicity for nontarget organisms, including man. A large body of evidence has been gathered over the last 20 years in laboratory animal studies showing that exposure to environmental chemicals, many of which are pesticides or pesticide-related, can produce immune dysfunction. This evidence has resulted from studies following acute or subchronic exposure regimens exposing animals to relatively high doses of test agents. The evidence has prompted the study of immune function in humans inadvertently exposed to some of these agents; see review (7). Although the results from several of these studies suggested that immune dysfunction had occurred, those of others were equivocal. In contrast to immune dysfunction, the most likely health consequences to man following exposure to pesticides may be respiratory tract allergies (e.g., asthma, rhinitis) or allergic contact dermatitis.

With respect to assessing health, there are several key issues associated with pesticide-induced immunomodulation which must be considered. This paper will evaluate the data base on pesticide-induced immunotoxicity, highlight the methodologies used to assess immune modulation, and discuss important issues associated with health assessment.

Potential for Adverse Effects of Pesticides on the Immune System

Exposure to pesticides can provoke a variety of immune reactions. These reactions can be classified into (a) modulation of normal immune responses (immune dysfunction), characteristically manifested as immunosuppression, and (b) pathological enhancement of the immune response, most often manifested as hypersensitivity or autoimmunity. The number of reviews on this subject underscores the interest in and concern for the potential of pesticides to alter immune function (8-15). The two general categories of immune alterations induced by pesticides are discussed below.

Immune Dysfunction. As summarized in Table I, several classes of pesticides alter immune function and resistance to infection in laboratory animals. Although sufficient evidence exists that pesticides affect immune function and host resistance in rodents, the data in humans are less clear. For example, even though hemato-logical changes were observed in humans exposed to pesticides in a greenhouse environment (where high concentrations of the agent are often present), there was no indication of altered immune status (38). However, in a study of pesticide workers exposed to combina-tions of four pesticides (malathion, parathion, DDT, and hexachloro-cyclohexane), 73% demonstrated alterations of serum immunoglobulin (Ig) levels (39). In another study, increases in serum IgG but decreases in serum IgM and the C-3 component of complement were reported to occur in 51 men exposed to chlorinated pesticides, as compared with a 28-man control group (40). In the latter two studies, in spite of immune changes, a direct association between increased susceptibility to infection with changes in immune status was absent.

Recent interest in the potential adverse effects of pesticides on the immune system has stemmed from studies in mice and humans exposed to low levels of aldicarb, a carbamate pesticide. These studies reported suppression of antibody responses following exposure for 34 days to levels of aldicarb as low as 1 ppb in drinking water (32). However, Thomas et al. (41), using a similar exposure regimen and mouse strain but a more comprehensive testing protocol, observed no aldicarb-related effects on immunity or susceptibility to infec-tion. In another study (33), women chronically ingesting low levels of aldicarb-contaminated groundwater had altered numbers of T-cells, including a decreased CD4:CD8 helper/suppressor cell ratio. However, these individuals did not demonstrate any increased incidence of infection associated with aldicarb exposure.

In addition to their active compounds, pesticide formulations often contain by-products of the manufacturing process and a quantity of inert ingredients. The potential contribution made by all known or potential additional components in any pesticide preparation must be considered in assessment of a pesticide for immunotoxic potential. For example, O,O,S-trimethyl phosphorothioate (OOS-TMP), a contami-nant of malathion, has been shown to alter immune function (19, 42-46). Mice exposed to OOS-TMP displayed reduced humoral responses, reduced cell-mediated immune responses, and altered macrophage function.

Little work has been done concerning the immunomodulatory effects of pesticides on the developing immune system; see review (47). In some recent studies, mice exposed to chlordane in utero displayed a decreased contact sensitivity response and an increased survival and antibody response to influenza as adults (25-28). Exposure of mice to this compound as adults, however, did not significantly alter cellular or humoral immune functions (48). These results suggest that the developing immune system may be more suscep-tible to immunomodulation by chlordane than the adult immune system, and are consistent with earlier studies (49, 50) demonstrating a greater sensitivity of mice to the effects of 2,3,7,8-tetrachloro-dibenzo-p-dioxin (TCDD) on developing than on adult immune systems.

Hypersensitivity. In addition to inducing immune dysfunction, pesticides have the potential to exert immunomodulatory effects through the induction of allergic hypersensitivity and autoimmune disease. Pesticide-related hypersensitivity reactions generally are

Table I. Examples of Pesticides Reported to Modulate Immunity

Pesticide	Species	Refs.	Effects
Organosphosphates			
Methylparathion	Rabbit	(16)	Thymus atrophy and reduced delayed-type hypersensitivity (DTH) response
	Mouse	(17)	Decreased resistance to infection with S. typhimurium
Parathion	Mouse	(18)	Increased mortality to parathion in cytomegalovirus infected mice
Malathion	Mouse	(19)	Suppression of cell-mediated
	Human	(20)	immunity, in vitro
Organochlorines			
DDT	Rabbit	(16)	Thymus atrophy and reduced DTH
Mirex	Chicken	(21)	Decreased IgG levels
Hexachlorobenzene	Mouse	(22)	Increased sensitivity to endotoxin and malaria
Dieldrin	Mouse	(23, 24)	Decreased humoral immunity and increased susceptibility to viral infection
Chlordane	Mouse	(25, 26)	Decreased contact hypersensitivity after in utero exposure
	Mouse	(27, 28)	Increased survival and antibody response to influenza after in utero exposure
	Mouse	(29)	In vitro suppression of humoral- and cell-mediated immune responses
	Human	(30)	Decreased T-lymphocyte proliferation

Continued on next page

Table I. Examples of Pesticides Reported to Modulate Immunity
(continued)

Pesticide	Species	Refs.	Effects
Chlorophenoxy Compounds			
2,4-dichloro-phenoxyacetic acid (2,4-D)	Mouse	(31)	Enhanced T- and B-cell responses following dermal application
Carbamates			
Carbofuran	Rabbit	(16)	Reduced DTH
	Mouse	(17, 9)	Bacterial infection with S. typhimurium
Aldicarb	Mouse	(32)	Decreased humoral immune response
	Human	(33)	Increased in vitro responses to Candida antigen, increased suppressor cells and decreased helper/suppressor cell ratio
Carbaryl	Human	(20)	Decreased T-lymphocyte proliferation in vitro
	Mouse	(34)	Increased serum immuno-globulin levels
Organotins			
Triphenyltin hydroxide (TPTH)	Rat	(35)	Reduced DTH, decreased T-cell response
Tributyltin oxide (TBTO)	Rat	(36)	Reduced cell-mediated, natural killer cell, and macrophage responses and decreased resistance to T. spiralis
Additives and Contaminants			
Butoxide	Human	(20)	Decreased T-lymphocyte proliferation
Dicresyl	Rat	(37)	Decreased resistance to E. coli and S. aureus

confined to two of the four hypersensitivity responses as defined by
Coombs and Gell (51)--namely, delayed-type (contact) and immediate
hypersensitivity. Contact hypersensitivity is a T-lymphocyte
mediated inflammatory response, the cutaneous clinical manifestation
of which is allergic contact dermatitis. Immediate hypersensitivity
is mediated primarily by IgE antibodies and mast cells, with clinical
manifestations including allergic rhinitis, asthma, and (in rare
instances), systemic anaphylaxis. Both hypersensitivity responses
require initial exposure to an allergen (sensitization) to induce the
immune response. A subsequent (challenge) exposure then elicits the
clinical manifestations (52).

A monograph on contact hypersensitivity (8) lists over 40
pesticides of various chemical classes which have been implicated by
case reports as causal agents of contact dermatitis in humans. Other
pesticides not included in Cronin's list have also been reported to
cause contact dermatitis episodes (53-55). In spite of the large
number of such case reports, the incidence of documented allergic
contact dermatitis to any particular pesticide is rare. For example,
Winter and Kurtz (56) evaluated various environmental factors affecting the incidence of skin rashes in California vineyard workers.
They reported the incidence of skin rash as 24 per 1,000 workers per
year, a figure considerably higher than the reported incidences for
the general (2.1 per 1000) or agricultural work force (8.6 per 1000).
Despite this high incidence, the authors report little correlation
with any pesticide exposure. Most skin rashes were associated with
exposure to high temperatures during thinning and harvesting opera-
tions. Several investigators, however, have speculated that the
actual occurrence of contact dermatitis is greater than the reported
incidence (8, 57, 58). This may be due to the isolated nature of
agricultural work (especially migrant workers), exposure to numerous
chemicals, and lack of diagnostic follow-up.

Reports that pesticide exposure causes the development of
immediate allergic reactions such as rhinitis, asthma, or anaphylaxis
are also difficult to confirm. Many patients with underlying
allergic disease present with exacerbated symptoms following exposure
to pesticides. Most investigators, however, consider such reactions
irritative rather than allergic, although isolated case reports (59)
suggest that allergic reactions to organophosphate pesticides do
occur.

Autoimmunity. Autoimmune diseases are disorders of immune regulation
in which several different factors (e.g. viral, genetic, hormonal,
environmental) may each play a role. Autoimmune diseases may belong
to any of the four Coombs and Gell classifications of hypersensitivity and include the production of autoantibodies, destructive
inflammatory cell infiltrates in various organs, and deposition of
immune complexes in vascular beds. Chemically induced autoimmunity
may result from any of several possible mechanisms. These include
the alteration or release of autoantigens, or the cross-reaction of
the chemical with autoantigens, or alternatively a direct effect on
the immune system via lymphocytes or macrophages (60).

Pesticide-induced disorders resembling autoimmunity have been
reported but are rare. The presence of anti-dieldrin IgG antibody in

serum and coating red cell membranes was demonstrated in an individual with immunohemolytic anemia as reported by Hamilton et al. (61).
In other cases of pesticide exposure, general toxicity--rather than
an autoimmune process--may have been responsible for several reports
of aplastic anemia in humans (62-64). However, the presence of antibodies in experimental animals exposed to DDT or malathion conjugates
(65) suggests that similar immunopathologic responses predisposing to
autoimmunity might occur in man. Whereas current mammalian toxicology data requirements for pesticide registration should allow the
detection of autoimmune-related hematological disorders and certain
organ-specific changes, other autoimmune responses may be undetected.
Furthermore, laboratory animal model systems for autoimmunity are not
currently well-developed.

Specific Issues Surrounding Pesticide-Induced Immunomodulation

Testing Methodologies Used. There have been legal challenges based
on the assumption that exposure to environmental or occupational
chemicals produces immunologic disease (66). Without well-defined
criteria for comparing immune function in normal and chemically
exposed individuals, the potential exists for misuse or misinter-
pretation of immunological data at these proceedings (67). The complex interactions among the various cells and tissues of the immune
system are advantageous to the host but confound studies assessing
the impact of pesticides on the immune system. To maximize testing
accuracy and the ability to make meaningful risk assessment decisions, one must evaluate the immune system at a variety of different
levels.

In experimental animals, several comprehensive approaches for
immunotoxicity assessment have been proposed (68, 69). One generally
accepted approach advocates using a systematic, tiered assessment in
normal, healthy, young adult animals (69). The first tier of assays
provides a screening mechanism for identifying potential immunomodulatory compounds. The methods focus on evaluating relevant
pathological, hematological, and anatomical parameters associated
with the immune system in addition to limited B- and T-cell function
tests (e.g., antibody plaque forming cell assays, mixed lymphocyte
culture response). The second tier provides information concerning
the biological significance of effects observed in the first tier as
well as elucidating the cellular or molecular mechanism of action of
the compound. This can be accomplished using in vivo host resistance
models in which animals are challenged with infectious agents or
transplantable tumor cells or more specific in vitro immune function
assays.

In the case of hypersensitivity testing, a number of guinea pig
test methods are widely used for predicting contact allergenicity in
humans exposed to pesticides or to other chemicals. A description of
these tests and their usefulness in predicting human contact hypersensitivity incidence has been reviewed by Andersen and Maibach (70).
Several human patch test methods are also utilized for predictive
skin sensitization testing of chemicals or formulations (71). Though
not required by regulatory agencies, data from such testing are
valuable in assessing sensitization risk to humans.

Public Perception of Risk. As a result of the AIDS pandemic, the public has become aware of the role of the immune system in host defense and recognizes the potential problems associated with a compromised immune system. Considering the limited scientific evidence documenting immunosuppression in humans following low-level environmental exposure to pesticides, the laboratory observations in animals should be interpreted with caution and the relevance to humans placed in its proper perspective. The inability to defini-tively correlate pesticide exposure in humans with adverse health effects due to immunosuppression may be the result of several factors. A principal factor is the difficulty in pinpointing the health consequences of minor immunological perturbations resulting from low pesticide exposure levels. In this case, the actual level of human exposure may be significantly lower relative to the experi-mental animal models, making extrapolation of animal data to humans difficult or impossible. Other contributing factors could include the difficulty in defining "normal" immune status in humans, and the uncertainties regarding loss of immune reserve and subsequent appearance of disease.

Status of Epidemiologic Studies. The usefulness of epidemiology in linking environmental exposure to pesticides or other chemicals with altered immune status and subsequent changes in resistance to diseases such as cancer remains questionable at present; see review (72). As stated above, variability among the population with respect to what is considered a normal immune response, the inability to objectively measure immune status, and confounding factors involved in interpretation of these measures decrease the probability of using epidemiology to identify pesticides that may be responsible for caus-ing immune injury. Subtle perturbations in immune function following exposure to pesticides may not, in every instance, result in a rele-vant health effect. Conceivably, these changes could increase the likelihood of adverse immune-related health effects only during the brief period when they occur. On the other hand, minor health changes caused by alterations in immune function may not be detected in an epidemiological survey. However, current epidemiological data concerning exposure to some of the more toxic environmental chemicals do not support a strong link between subtle immunological perturba-tions and biologically relevant changes in resistance to disease. In spite of this, it cannot be ruled out that inherent problems asso-ciated with studies of this nature have prevented the detection of evidence for this link.

    Contact hypersensitivity accounts for a significant percentage of occupational adverse skin reactions and, though case reports are rare, is thought to be an important cause of skin reactions in pesticide-exposed workers (8). The epidemiological data must be improved, however, to confirm this association (12) since other causal factors may be involved (56).

Animal-to-Human Extrapolation. Current data bases containing human and experimental animal data provide some interspecies correlation in comparisons of immune perturbations caused by exposure to pesticides and other chemicals. The same approaches used by toxicologists with other target organs for extrapolation from experimental animals to man are valid for the immune system. For example, the first choice of an experimental animal is one in which the pesticide is absorbed, distributed, biotransformed, and excreted in a manner similar to man. The next choice for an experimental animal is one in which the desired endpoint can best be measured. The rodent has served as the principal animal model for man in delineating immune processes. Furthermore, the majority of compounds known to directly modulate the immune system in man do so in a fashion similar to the mouse and, based upon available data, also to the rat. When no information on pharmacokinetics of the pesticide exists for humans, its laboratory evaluation in more than one nonhuman species increases the likelihood of accurately predicting its immunotoxicological effects in humans.

In the case of contact hypersensitivity, some difficulty arises in extrapolating from the animal sensitization test data to human sensitization risk. This is due to the fact that the published animal test data (primarily guinea pig maximization test results) generally indicate a sensitization potential far greater than the actual human experience would indicate. Lastly, methods for identifying compounds capable of autoimmune responses exist but have not yet been validated.

Conclusions

The discipline of immunotoxicology represents a relatively new area of toxicology. As a result, only a limited data base exists for pesticides which have been adequately examined in laboratory and epidemiological studies. However, based on the limited studies conducted in rodents, selected pesticides or their by-products can adversely affect the immune system through mechanisms which may include disruption of cell maturation, regulation, or cytotoxic processes, thus leading to altered host resistance and possible cancer. The animal data, along with the current knowledge about the pathogenesis of diseases associated with immune dysregulation, suggest that these and certain other pesticides (or related compounds) may affect the immune system in humans. With the exception of limited in vitro exposure studies and clinical data demonstrating that certain pesticides induce hypersensitivity, no substantial evidence as yet exists that exposure to pesticides, either in the workplace or through casual contact, induces significant immune dysfunction in humans.

Acknowledgments

The authors wish to acknowledge input from Drs. N. Kerkvliet, M. Luster, A. Munson, M. Murray, D. Roberts, J. Silkworth, and R. Smialowicz in the preparation of this manuscript.

Literature Cited

1. Heise, E. R. Environ. Health Perspect. 1982, 43, 9-19.
2. Penn, I. In Immunotoxicology and Immunopharmacology; Dean, J. H.; Luster, M. I.; Munson, A. E.; Amos, H., Eds.; Raven Press: New York, 1985, pp 79-89.
3. Gottlieb, M.; Schroff, R.; Shanker, H.; et al. New Engl. J. Med. 1981, 305, 1425-1431.

4. Descotes, J. Immunotoxicology of Drugs and Chemicals; Elsevier:
    Amsterdam, 1986; 400 pp.
5. Stjernsward, J. J. Natl. Cancer Inst. 1966, 36, 1189-1195.
6. Ward, E. C.; Murray, M. J.; Dean, J. H. In Immunotoxicology and
    Immunopharmacology, Dean, J. H.; Luster, M. I.; Munson, A. E.;
    Amos, H. E., Eds.; Raven Press: New York, 1985; pp 291-303.
7. Dean, J. H.; Thurmond, L. M. Toxicol. Pathol. 1987, 15, 265-
    271.
8. Cronin, E. Contact Dermatitis; Churchill Livingstone:
    Edinburgh, 1980; p 399.
9. Street, J. C. In CRC Immunologic Considerations in Toxicology,
    Vol. I; Sharma, R. P., Ed.; CRC Press, Inc.: Boca Raton, Fla.,
    1981; pp 45-66.
10. Dean, J. H.; Lauer, L. D. In The Basic Science of Poisons;
    Third Edition; Macmillan: New York, 1984; pp 245-285.
11. Caspritz, G.; Hadden, J. Toxicol. Pathol. 1987, 15, 320-332.
12. Edmiston, S.; Maddy, K. T. Vet. Hum. Toxicol. 1987, 29, 391-
    397.
13. Exon, J. H.; Kerkvliet, N. I.; Talcott, P. A. Environ.
    Carcinogen Rev. (J. Environ. Health) 1987, C5, 73-120.
14. Luster, M. I.; Blank, J. A.; Dean, J. H. Ann. Rev. Pharmacol.
    Toxicol. 1987, 27, 23-49.
15. Sharma, R. P. Toxicol. Ind. Health 1988, 4, 373-380.
16. Street, J. C.; Sharma, R. P. Toxicol. Appl. Pharmacol. 1975,
    32, 587-602.
17. Fan, A.; Street, J. C.; Nelson, R. M. Toxicol. Appl. Pharmacol.
    1978, 45, 235.
18. Selgrade, M. K.; Daniels, M. J.; Illing, J. W.; Ralston, A. L.;
    Grady, M. A.; Charlet, E.; Graham, J. A. Toxicol. Appl.
    Pharmacol. 1984, 76, 356-364.
19. Rodgers, K. E.; Grayson, M. H.; Imamura, T.; Devens, B. H.
    Pesticide Biochem. Physiol. 1985, 24, 260-266.
20. Lee T. P.; Moscali, R.; Park, B. H. Res. Commun. Chem. Pathol.
    Pharmacol. 1979, 23, 597.
21. Rao, D.S.V.S.; Glick, B. Proc. Soc. Exp. Biol. Med. 1977, 154,
    27-29.
22. Loose, L. D.; Silkworth, J. B.; Benitz, K. F.; Mueller, W.
    Infect. Immun. 1978, 20, 30-35.
23. Bernier, J.; Hugo, P.; Krzystyniak, K.; Fournier, M. Toxicol.
    Lett. 1987, 35, 231-240.
24. Krzystyniak, K.; Hugo, P.; Flipo, D.; Fournier, M. Toxicol.
    Appl. Pharmacol. 1985, 80, 397-408.
25. Spyker-Cranmer, J. M.; Barnett, J. B.; Avery, D. L.; Cranmer, M.
    F. Toxicol. Appl. Pharmacol. 1982, 62, 402-408.
26. Barnett, J. B.; Soderberg, L.S.F.; Menna, J. H. Toxicol. Lett.
    1985, 25, 173-183.
27. Menna, J. H.; Barnett, J. B.; Soderberg, L.S.F. Toxicol. Lett.
    1985, 24, 45-52.
28. Barnett, J. B.; Holcomb, D.; Menna, J. H.; Soderberg, L.S.F.
    Toxicol. Lett. 1985, 25, 229-238.
29. Johnson, K. W.; Kaminski, N. E.; Munson, A. E. J. Toxicol.
    Environ. Health 1987, 22, 497-515.
30. Roux, F.; Treich, I.; Brun, C.; Desoize, B.; Fournier, E.
    Biochem. Pharmacol. 1979, 28, 2419.

31. Blakely, B. R.; Schiefer, B. H. J. Appl. Toxicol. 1986, 6, 291-295.
32. Olson, L. J.; Erickson, B. J.; Hinsdill, R. D.; Wyman, J. A.; Porter, W. P.; Binning, L. K.; Bidgood, R. C.; Nordheim, E. V. Arch. Environ. Contam. Toxicol. 1987, 16, 433-439.
33. Fiore, M. C.; Anderson, H. A.; Hong, R.; Golubjatnikov, R.; Seiser, J. E.; Nordstrom, D.; Hanrahan, L.; Belluck, D. Environ. Res. 1986, 41, 633-645.
34. Andre, F.; Gillon, F.; Andre, C.; Lafont, S.; Jourdan, G. Environ. Res. 1983, 32, 145.
35. Vos, J. G.; Van Logten, M. J.; Kreeftenberg, J. G.; Kruizinga, W. Toxicologist 1984, 29, 325-336.
36. Vos, J. G.; de Klerk, A.; Krajnc, E. I.; Kruizinga, W.; van Ommen, B.; Rozing, J. Toxicol. Appl. Pharmacol. 1984, 75, 387-408.
37. Olefir, A. I. Vrach. Delo. 1973, 8, 137.
38. Kundiev, Y. I.; Krasnyuk, E. P.; Viter, V. P. Toxicol. Lett. 1986, 33, 85-89.
39. Kashyap, S. K. Toxicol. Lett. 1986, 33, 107-114.
40. Wysocki, J.; Kalina, Z.; Owczarzy, I. Med. Pract. 1985, 36, 111-117.
41. Thomas, P. T.; Ratajczak, H. V.; Eisenberg, W. C.; Furedi-Machacek, M.; Ketels, K. V.; Barbera, P. W. Fundam. Appl. Toxicol. 1987, 9, 82-89.
42. Devens, B. H.; Grayson, M. H.; Imamura, T.; Rodgers, K. E. Pesticide Biochem. Physiol. 1985, 24, 251-259.
43. Rodgers, K. E.; Imamura, T.; Devens, B. H. Immunopharmacology 1985, 10, 171-180.
44. Rodgers, K. E.; Imamura, T.; Devens, B. H. Toxicol. Appl. Pharmacol. 1985, 81, 310-318.
45. Rodgers, K. E.; Imamura, T.; Devens, B. H. Immunopharmacology 1986, 12, 193-202.
46. Rodgers, K. E.; Imamura, T.; Devens, B. H. Toxicol. Appl. Pharmacol. 1987, 88, 279-281.
47. Thomas, P. T.; Faith, R. W. In Immunotoxicology and Immunopharmacology, Dean, J. H.; Luster, M. I.; Munson, A. E.; Amos, H. E., Eds.; Raven Press: New York, 1985; pp 305-313.
48. Johnson, K. W.; Holsapple, M. P.; Munson, A. E. Fundam. Appl. Toxicol. 1986, 6, 317-326.
49. Moore, J. A.; Faith, R. E. Environ. Health Perspect. 1976, 18, 125-131.
50. Thomas, P. T.; Hinsdill, R. D. Drug Chem. Toxicol. 1979, 2(142), 77-98.
51. Coombs, R.R.A.; Gell, P.G.H. In Clinical Aspects of Immunology; Gell P.G.H.; Coombs, R.P.A., Eds.; Blackwell Scientific Publishers: Oxford, 1968; Chapter 20.
52. Stanworth, D. R. In Immunotoxicology and Immunopharmacology; Dean, J. A.; Luster, M. I.; Munson, A. E.; Amos, H. E., Eds.; Raven Press: New York, 1985 pp 91-98.
53. Johnsson, M.; Buhagen, M.; Leira, H. L.; Solvang, S. Contact Dermatitis 1983, 9, 285-288.
54. Mathias, C.G.T.; Andersen, K. E.; Hamann, K. Contact Dermatitis, 1983, 9, 507-509.
55. Meding, B. Contact Dermatitis 1986, 15, 187.

56. Winter, C. K.; Kurtz, P. H. <u>Bull. Environ. Contam. Toxicol.</u> 1985, <u>35</u>, 418-426.
57. Matsushita, T.; Aoyama, K. <u>Ind. Health</u> 1980, <u>18</u>, 31-39.
58. Hogan, D. J.; Lane, P. R. <u>Can. Med. Assoc. J.</u> 1985, <u>132</u>, 387-389.
59. Bryant, D. H. <u>Aust. N.Z. J. Med.</u> 1985, <u>15</u>, 66-68.
60. Bigazzi, P.E. In <u>Immunotoxicology and Immunopharmacology</u>; Dean, J. H.; Luster, M. I.; Munson, A. E.; Amos, H. A., Eds.; Raven Press: New York, 1985; pp 277-290.
61. Hamilton, H. E.; Morgan, D. P.; Simmons, A. <u>Environ. Res.</u> 1978, <u>17</u>, 155-164.
62. Roberts, J. H. <u>South. Med. J.</u> 1963, <u>56</u>, 632-635.
63. Sanchez-Medal, L.; Castanedo, J. P.; Garcia-Rojas, F. <u>N. Engl. J. Med.</u> 1963, <u>269</u>, 1365-1367.
64. Loge, J. P. <u>J. Am. Med. Assoc.</u> 1965, <u>193</u>, 110-114.
65. Centeno, E. R.; Johnson, W. J.; Sehon, A. H. <u>Int. Arch. Aller.</u> 1970, <u>37</u>, 1-13.
66. Elam, Betty et al. vs. Alcolac, Inc., et al., No. CV 81-26972, Circuit Court of Jackson County, Missouri, 1985.
67. Roisman, A. In <u>ICET Symposium III: Immunotoxicology from Lab to Law</u>; Institute for Comparative and Environmental Toxicology, Cornell Univesity: Ithaca, N.Y., 1988; pp 105-114.
68. Exon, J. H.; Koller, L. D.; Talcott, P. A.; O'Reilly, C. A.; Henningsen, G. M. <u>Fundam. Appl. Toxicol.</u> 1986, <u>7</u>, 387-397.
69. Luster, M. I.; Munson, A. E.; Thomas, P. T.; Holsapple, M. P.; Fenters, J. D.; White, K. L., Jr.; Lauer, L. D.; Germolec, D. R.; Rosenthal, G. J.; Dean, J. H. <u>Fundam. Appl. Toxicol.</u> 1988, <u>10</u>, 2-19.
70. Andersen, K. E.; Maibach, H. I. In <u>Contact Allergy Predictive Tests in Guinea Pigs</u>; Andersen, K. E.; Maibach, H.I., Eds.; Karger: Basel, 1985; pp 263-290.
71. Stotts, J. In <u>Current Concepts in Cutaneous Toxicity</u>; Drill, V. A.; Lazar, P., Eds.; Academic Press: New York, 1980; pp 41-53.
72. Ellis, E. In <u>ICET Symposium III: Immunotoxicology from Lab to Law</u>; Institute for Comparative and Environmental Toxicology, Cornell University: Ithaca, N.Y., 1988; pp 37-45.

RECEIVED August 1, 1989

Chapter 7

Dietary Inhibition of Cancer

Diane F. Birt

Eppley Institute for Research in Cancer, University of Nebraska Medical
Center, 42nd Street and Dewey Avenue, Omaha, NE 68105–1065

There is little data on the inhibition of pesticide
carcinogenicity by diet; however, it is anticipated
that some of the mechanisms which will be identified
for the interactions between nutrition and other
dietary components will also apply to the interaction
between dietary components and pesticides. Both
nutrient and non-nutrient dietary factors have been
found to modify chemical carcinogenesis. Nutrients
generally act to alter the process of carcinogenesis
and do not function as initiators or promoters of
cancer. Current research on macronutrients is assessing the interactions between dietary fat, calories and
fiber in cancer induction and promotion. The most
promising leads for micronutrient inhibition of cancer
are with vitamins A, C and E and with selenium. Non-
nutrient dietary factors are of particular interest
because of negative associations between certain fruit
and vegetable consumption and cancer rates. Examples
of such factors include some flavonoids and terpenes.
This paper will provide examples of how dietary
components may function in the inhibition of cancer.

Dietary factors have been investigated for their involvement in
cancer etiology ever since the importance of environmental factors
in cancer became apparent. However, an understanding of the involvement of dietary factors in the cancer process has developed
very slowly. A prime reason for this slow progress is probably the
elusive nature of the influence of diet in cancer etiology. For
example, in comparison with the association between cigarette smoking and cancer, the association between diet and cancer is very
weak. In considering different sites of cancer, dietary components
often have conflicting effects, whereas smoking is associated with
an increased rate of cancer at every site where it has an influence.
However, diet is probably one of the most important means whereby
nonsmokers can control their cancer risk.

0097–6156/89/0414–0107$06.00/0
© 1989 American Chemical Society

It is certainly reasonable to assume that dietary factors which
have been studied for their ability to prevent chemically-induced
cancer may be useful in the prevention of cancer induced by pesticides; however, I could not find recent studies which have explored
the relationship between dietary factors and cancer induced by
pesticides. The absence of this important data in the literature is
probably due to the decision on the part of investigators studying
the relationship of diet and cancer to use well-established, well-
defined models of chemical carcinogenesis for their dietary studies.
I anticipate that diet will be used to modulate pesticide carcinogenesis as our understanding of pesticides and their relationship to
cancer expands. Furthermore, I am not aware of pesticides being
studied as potential dietary inhibitors of cancer, although, as
pointed out in the presentation by Dr. Gary Williams, pesticides
such as DDT can influence the metabolism of certain carcinogens in a
manner which would reduce the carcinogenicity of these agents.

This chapter will provide an overview of the factors which have
been studied most extensively as potential inhibitors of cancer. It
will stress data in the areas where recommendations have been made
to the public (1,2). We will begin with a discussion of the macronutrients which have received the greatest attention for their potential for modifying cancer risk, dietary fat and fiber, and the
relation of these nutrients to calorie intake. I will then describe
data suggesting that micronutrients such as vitamins A, C, and E,
and the trace element selenium, may have some potential in the inhibition of cancer. Finally, I will present some recent data in
support of cancer prevention by some non-nutrient components of
fruits and vegetables.

Dietary Fat, Fiber and Calories

High fat diets are generally associated with an increased risk of
cancer in people consuming such diets and an enhancement of carcinogenesis in animals fed such diets. The converse of this should also
certainly be considered, a reduction in cancer risk in people consuming low fat diets and an inhibition of cancer in animals fed low
fat diets. In the massive compilation of data prepared by the
National Research Council (NRC) on Diet, Nutrition and Cancer (1),
the number of studies showing a relationship between dietary fat and
cancer was more impressive than the relationship between diet and
any other nutrient. "The committee concluded that of all the dietary components it studied, the combined epidemiological and experimental evidence is most suggestive for a causal relationship between
fat intake and the occurrence of cancer".

People consuming low fat diets have been found to have lower
rates of colon, breast, pancreas, and prostate cancer (1,2).
Animals consuming low fat diets generally had reduced rates of
cancers induced in the breast, colon, liver, pancreas, lung and skin
as shown in Table I [previously reviewed by Birt, (3)]. This is
particularly true with diets containing fats high in W-6 fatty
acids. Recent work indicates possible inhibition of cancer by W-3
fatty acids. The primary difficulty in interpreting the studies of
the effects of dietary fat on carcinogenesis is the problem of
separating the effects of fat from the effects of calories. Diets

Table I.   Summary of the Effects of Dietary Fat on Tumorigenesis

Site	Animal	Agent	Effect of high fat diet[*]
Skin	Mouse	None (spontaneous)	↑
		Polycyclic hydrocarbon	↑
		UV light	↑
Lung	Mouse	None	↑,NE
	Hamster	Benzo(α)pyrene (BP) on ferric oxide	↑
		N-nitrosobis(2-oxopropyl)amine (BOP)	↑
Mammary gland	Rat	$\overline{7}$,12-dimethylbenz(a)anthracene	↑
	Mice	Spontaneous	↑
	Rat	Acetylaminofluorene (AAF)	↑
	Rat	Methylnitrosourea (MNU)	↑
	Rat	X-irradiation	↑
Colon	Rat	1,2-dimethylhydrazine (DMH)	↑,NE
		Azoxymethane (AOM)	↑
		methylazoxymethanol acetate (MAM)	↑,NE
		3,2'-dimethyl-4-amino-biphenyl (DMAB)	↑
		MNU	↑
Liver	Mouse	None	↑
	Rat	Aminoazo dyes	↑
	Hamster	Aflatoxin B_1 (AFB)	↑
		AAF	↑
		BOP	↑
Pancreas	Hamster	BOP	↑
	Rat	Azaserine (AZA)	↑
Prostate	Rat	None (promoted with testosterone)	↑
Brain	Mouse	None	↑
Ear duct	Rat	DMH	NE
Kidney	Rat	DMH	E
	Hamster	BOP	↑

[*] NE = no effect.   Adapted from reference 3.

enriched in fat have an elevated caloric density. Thus, animals
consuming these diets may consume extra calories, or they may
utilize the calories consumed in a more efficient manner.

This is important in understanding the influence of calories on
cancer because it is well known that reduced calorie intake can also
inhibit carcinogenesis (3) and calorie intake appears to be a factor
in human breast and colon cancer (4,5). Studies using the rat
mammary carcinogenesis system indicated that restriction of a high
fat diet to the level of energy consumption by the low fat control
resulted in considerable inhibition of tumorigenesis, suggesting
that some of the influence of dietary fat on mammary tumorigenesis
may be due to the influence of calories (6). However, recent investigations in my laboratory demonstrated that pancreatic carcinogenesis was enhanced in hamsters fed a high fat diet irrespective of
whether diets were fed at constant and slightly restricted caloric
amounts or in an ad libitum manner where excessive calories were
consumed by the high fat group (7).

Dietary fiber's effects on carcinogenesis have been the topic
of much debate (8). The public widely believes that the consumption
of fiber results in reduced risk of colon cancer, yet data from case
control studies with humans indicates that fiber consumption is not
strongly associated with reduced risk of colon cancer, and some
studies have even shown an increased risk associated with high fiber
intakes (9). Furthermore, data from studies with animals suggest
that fiber can inhibit, enhance or have no influence on colon carcinogenesis, depending upon the form of fiber given, when it is
given relative to the carcinogen, and the type of diet in which it
is given (8).

Vitamin and Mineral Inhibition of Cancer

Vitamin A has been one of the most extensively studied micronutrients for the prevention of cancer (10). The nutritional role of
vitamin A includes regulation of normal differentiation. Cancer
epidemiology suggested that the consumption of foods rich in vitamin
A may be associated with a reduced rate of a number of forms of
cancer, including lung (11-14). Such results further suggest that
vitamin A or its precursors may be particularly beneficial in the
prevention of lung cancer in smokers (58). Animal investigations
have used a number of analogues of vitamin A which possess reduced
toxicity and were anticipated to provide increased protection
against cancer as well as using vitamin A itself. A summary of
studies with vitamin A deficiency increasing cancer induction and
supplementation with vitamin A or its analogues (retinoids) inhibiting cancer induction was recently published (10), and the
overall effects are shown in Table II. In general, vitamin A and
analogues of vitamin A were particularly effective in preventing
urinary bladder tumorigenesis in a model using a nitrosamine carcinogen (15), but not in another model using a nitrofuran (16).
Mammary gland carcinogenesis in rats was inhibited by vitamin A and
its analogues (17), but mammary tumorigenesis in mice was not influenced (18). In general, retinoids have shown variable effects on
tumorigenesis in the skin, lung, liver, colon, and pancreas (10).
More recent studies with retinoids have assessed the potential of

Table II. Summary of the Influence of Retinoids on
Chemical Carcinogenesis[1]

Site	Species	Agent[2]	Retinoid treatment[3]	Effect[3]
Skin	Mice	DMBA	Vit. A analogues	↓
	Mice	DMBA + TPA	Retinoic acid	↑
	Mice	UV light	Retinoic Ac	↑
Salivary	Hamsters	DMBA	Vit. A def.	NE
glands		DMBA	13 cis ret. Ac	↓
Mammary	Rats	DMBA	Ret. acetate	↓
gland	Rats	MNU	Ret. acetate + analogues	↓
	Mice	DMBA	analogues	NE
	Mice	None	Retinyl acetate	NE
Forestomach	Hamsters	DMBA	Vit. A palmitate	↓
	Hamsters	BP	Vit. A palmitate	↓
Urinary	Rats	BBN	Vit. A analogues	↓
bladder	Mice	BBN	Vit. A + analogues	↓
	Rats	MNU	13 cis ret. Ac.	↓
	Rats	FANFT	Vit. A def.	↑
	Rats	FANFT	Ret. palmitate	NE
Lung	Rats	3MC	Vit. A def.	↑
	Rats	3MC	Ret. acetate	NE
	Hamsters	BP-FeO$_3$	Retinyl acetate	↑
Trachea	Hamsters	MNU	Vit. A analogues	↑
Liver	Rats	AFB$_1$	Vit. A def.	↑
	Rats	DMAB	Ret. acid	↓
	Hamsters	BOP	Vit. A analogues	↑
	Rats	AZA	Vit. A analogues	↑ NE
Colon	Rats	AFB$_1$	Vit. A def.	↑ NE
	Rats	MNNG	Vit. A def.	↓
	Rats	DMH	13 cis ret. Ac	↓
	Rats	MNU	Vit. A analogues	↑ NE
Pancreas	Hamsters	BOP	Vit. A analogues	↑ NE
	Rats	AZA	Vit. A analogues	↓

[1] References were published previously by Birt (10).
[2] Abbreviations are as in Table I and: DMBA = 7,12-dimethylbenz-
anthracene; TPA = 12-0-tetradecanoylphorbol-13-acetate; BBN =
butyl(4-hydroxy)butyl-nitrosamine; FANFT = N-[4-(5-nitro-2-furyl)-2-
thiazolyl] formamide; 3MC = 3-methylcholanthrene; MNNG = N-methyl-
N'-nitro-N-nitrosoguanidine.
[3] Def. = deficiency; ret. Ac. = retinoic acid; NE = negative.

combining retinoids with other dietary agents such as selenium,
vitamin E, or vitamin C which may prevent cancer. Such studies have
had mixed results. For example, studies by Ip indicated a
potentiation of the inhibition of cancer in rats treated with
selenium and vitamin A following carcinogen treatment (19), but
results from our laboratory indicated an inhibition of skin tumorigenesis in mice given the retinoid 4-hydroxy-retinamide but an abla-
tion of this effect when selenium or selenium and vitamin E were
added to the diet (Pelling et al., unpublished observation).
     Ascorbic acid has long been known to prevent nitrosation (20).
It is quite possible that dietary ascorbic acid is important in
preventing human cancer which is caused by endogenous nitrosamine
formation (21). There is evidence, for example, that gastric cancer
risk, which is a form of cancer likely to be caused by endogenous
nitrosamine formation under some conditions, is lower in areas where
the consumption of fresh fruits and vegetables is high (21). Animal
investigations clearly demonstrated that nitrosamine formation from
nitrite and morpholine was reduced in the presence of ascorbic acid
and further that cancer induced by giving nitrite and morpholine was
inhibited in animals given ascorbic acid (21). In addition, ascorbic acid has been shown to inhibit cancer induced by a number of
preformed carcinogens as reviewed previously (10) and shown in Table
III.
     Alpha-tocopherol has been shown to be useful in the prevention
of nitrosation in non-aqueous systems (20,22). Additionally, carcinogenesis by preformed carcinogens has been inhibited by alpha-
tocopherol, as is shown in Table IV. Interestingly, 7,12-dimethyl-
benz(a)anthracene (DMBA)-induced carcinogenesis in the mammary gland
was inhibited only in rats fed high fat or high selenium levels with
the vitamin E supplement (23,24).
     Selenium has been extensively studied for its ability to
inhibit carcinogenesis (Table V). Epidemiological studies have
suggested that elevated selenium in the blood was associated with a
reduction in cancer rates at a number of sites (25). Investigations
using animal models have demonstrated a consistent inhibition of
mammary and colon carcinogenesis in rats, but enhancement of carcinogenesis was reported in the pancreas and skin of animals fed
high selenium diets and treated with chemical carcinogens (10,26).
Pancreatic carcinogenesis of ductular morphology, similar to the
most common forms of the human disease, was elevated only under
specific conditions, whereas pancreatic acinar cell nodules were
inhibited (26,27), suggesting possible inhibition of acinar pancreatic tumorigenesis. Using the two-stage model of skin tumorigenesis, we found an enhancement of skin tumor promotion in mice fed
dietary selenium supplements following a high dose of carcinogen.
With lower doses of carcinogen no influence was observed (Pelling et
al., unpublished data).

Non-Nutrient, Inhibitors of Cancer from Fruits and Vegetables

The consumption of fruits and vegetables has been associated with a
reduction in cancer rate at a number of sites. Studies have reported reduced rates of gastric, colon and rectal, bladder, prostate and
breast cancer rates in association with cruciferous vegetables as

Table III. Summary of the Influence of Vitamin C on Chemical
Carcinogenesis[1]

Site	Species	Agent[2]	Vitamin C route (dosage)	Effects
Skin	Mouse	UV	Diet (100 g/kg)	↓
	Guinea pig	3MC	Varied	NE
	Guinea pig	3MC	ip	↑
Nose	Hamster	DEN + cigarette smoke	Diet (10 g/kg)	↓
Trachea	Hamster	DEN + cigarette smoke	Diet (10 g/kg)	↓
Lung	Mouse	DMN	Diet (23 g/kg)	↓
Mammary	Rat	DMBA	Water (2.5 g/kg)	NE
Colon	Rat	DMH	Diet (2.5–10 g/kg)	↓
	Rat	DMH	Diet (50 g/kg)	↑
	Mouse	DMH	Diet (<0.5 g/kg)	↓
	Rat	MNU	Diet (2.5–10 g/kg)	NE
Kidney	Rat	DMH	Diet (2–10 g/kg)	↓
	Hamster	estradiol		↓
	Hamster	DES		↓
Urinary bladder	Rat	BBN	Diet (50 g/kg) Ascorbic acid	NE
		BBN	Diet (50 g/kg) Sodium ascorbate	↑
		MNU	Diet (50 g/kg) Sodium ascorbate	↑

[1]References were published previously by Birt (10).
[2]Abbreviations are as in previous tables and: DEN =
diethylnitrosamine; DMN = dimethylnitrosamine.

Table IV. Summary of the Influence of Vitamin E on
Chemical Carcinogenesis[1]

Site	Species	Agent[2]	Vitamin E route (dosage)	Effect
Skin	Mouse	DMBA + TPA	Topical (17 mg 2 x wk)	↓
Skin	Mouse	DBP	Diet (25–50 g/kg)	NE
Cheek pouch	Hamster	DMBA	Oral (7 IV 2 x wk)	↓
			Oral (10 IV 2 x wk)	↓
Forestomach	Mouse	DMBA	Diet (10 g/kg)	↓
Colon	Mouse	DMH	Diet (0.6 g/kg)	↓
	Mouse	DMH	Diet (40 g/kg)	↑
	Rat	DMH	Vit. E def.	↑ or ↓
Mammary gland	Rat	DMBA	ig (unclear)	↓
		DMBA	Diet (30 mg/kg)	NE (low fat) ↓ (high fat)
		DMBA	Diet (50 mg/kg)	NE (low Se) ↓ (high Se)
		DMBA	Diet (2 g/kg)	NE

[1] References were published previously by Birt (10).
[2] Abbreviations are as in previous tables and: $\overline{DBP}$ = 3,4,9,10-dibenzpyrene.

Table V. Summary of the Influence of Selenium
on Chemical Carcinogenesis[1]

Site	Species	Agent[2]	Effect
Skin	Mice	uv light	↓
	Mice	3MC	↓
	Mice	α-Pyrene	↓
	Mice	DMBA + TPA	↑
Liver	Rat	DMAB	↓
	Rat	AAF	↓
	Rat	AFB	↑
Trachea	Hamster	MNU	NE
Lung	Rat	BOP	↓
Mammary gland	Mice	DMBA	↓
	Rat	DMBA	↓
	Rat	MNU	↓
Colon	Rat	DMBA	↓
	Rat	BOP	↓
	Rat	AOM	↓
Pancreas	Hamster	BOP	↑

[1] References were published previously by Birt (10).
[2] Abbreviations are as in previous tables.

reviewed previously (28). A recent case control study in China
found an elevated rate of stomach cancer among people consuming less
Chinese cabbage (29). Animal experiments in which dried cruciferous
vegetables are incorporated into diets have provided mixed support
for the potential of these vegetables in cancer prevention (reviewed
in 28). Studies of mammary (30) and liver (31) carcinogenesis and
lung metastasis (32) indicated an inhibition; however, studies of
skin (33), pancreas (33) and colon (34) cancer found an enhancement
of cancer under some experimental conditions. Particular compounds
reported to be present in or derived from those present in cruciferous vegetables, have been studied in recent years for their
ability to prevent carcinogenesis. Indole-3-carbinol has been
extensively studied, and although there is evidence that this agent
may be effective in inhibiting cancer initiation (35) it appears
that it has the potential to increase cancer rates if fed during
tumor promotion (36). Other components of cruciferous vegetables,
including dithiolthiones (37) and insoluble fibers (Birt,
unpublished results) are under active investigation.

Since studies on particular foods indicate varying degrees of
benefit in the prevention of cancer, many investigators in this area
are studying a wide variety of natural dietary components which do
not have known roles as essential nutrients but which have been
identified in foods as potential inhibitors of cancer. In the
limited space available here, I will provide an overview of some
selected agents and the types of studies being conducted on these
compounds.

Flavonoids. Flavonoids are ubiquitously distributed in vascular
plants. Over 2000 members of the flavonoid class have been described (38). They serve many functions in the plant, including
their activities as potent antioxidants and metal chelators (39).
Because of their wide distribution in plants, they are consumed in
large quantities by people. It is estimated that we consume at
least 1 gram (g) of plant flavonoid daily (40). Little is known
regarding the biological effects of plant flavonoids in mammalian
systems. An extensive survey of the mutagenicity of several
flavonoids has been conducted and two common plant flavonoids,
quercetin and kaempferol, were identified as being mutagenic (41).
Because of this observation, quercetin, the most mutagenic of the
two compounds, has been extensively studied for its carcinogenicity
(42-47). Only one study has shown any carcinogenicity, and this
report is suspect because other carcinogenic agents were used in the
experiment and the yield of cancer in the quercetin group was very
high (42).

More recently, quercetin has been investigated for its ability
to prevent the promotion phase of carcinogenesis. Nishino et al.
(48) topically applied DMBA to the backs of mice and teleocidin, the
promoter, was administered twice weekly beginning one week after
initiation. Quercetin was applied either 40 minutes before the
teleocidin or concurrently. Potent inhibition of the tumor incidence and prolongation of the latency period were observed after
administration of the flavonoid. For example, after 20 weeks of
promotion, an 83% reduction in the number of papillomas per mouse
was observed. In investigations where the flavonoid was adminis-

tered in the diet and skin tumorigenesis was investigated, no inhibition was reported (49). However, a recent report on mammary carcinogenesis induced by DMBA showed an inhibition with quercetin administered in the diet at a 1.0% level (50).

Numerous laboratories have investigated the influence of flavonoids on biochemical events which respond to tumor promoters (51,52). For example, a well known effect of tumor promoters applied to the skin of mice is the induction of ornithine decarboxylase (ODC). ODC is the rate limiting enzyme in polyamine synthesis and its activity is elevated by agents which lead to cellular proliferation. Several laboratories have investigated the influence of flavonoids applied before or with tumor promoters on the induction of ODC. The promoters cause a substantial increase in ODC and the flavonoids quercetin, kaempferol, luteolin, morin, fisetin, apigenin and robinetin (49,53,54) have been shown to inhibit this induction to varying extents.

The influence of flavonoids on the incorporation of ^{32}P into phospholipids, another hallmark of tumor promotion, has also been determined. Quercetin blocked teleocidin-stimulated ^{32}P-incorporation in a dose responsive manner (48). Finally, flavonoids were also effective in blocking promoter induced protein kinase C (PKC) induction as measured by the phosphorylation of H1 histone or the phosphorylation of endogenous proteins (55).

Studies from this laboratory determined the influence of apigenin on tumor promotion by pre-treating mice with apigenin prior to treatment with 12-0-tetradecanoylphorbol-13-acetate (TPA) and preliminary results are shown in Table VI (Wei et al., unpublished data). A dose response inhibition was observed in the incidence, and number of skin tumors and carcinomas were inhibited by both doses of apigenin. We have also determined that apigenin treatment inhibits the TPA-induced phosphorylation of epidermal proteins. Our aim is to develop a system to improve the uptake of apigenin to body organs and determine the influence of apigenin on tumorigenesis at other sites.

Table VI.   Inhibition of TPA-Induced Promotion by Apigenin in SENCAR Mice Initiated with DMBA[1]

Apigenin dose (μmole)	Number of effective mice	% Papilloma incidence	Number of papillomas/ effective mouse	% Carcinoma incidence	Number of carcinomas/ effective mouse
0	30	93	7.5	53	1.3
5	29	59	2.5	10	0.5
20	28	39	1.8	14	0.3

[1]Unpublished data (Wei, H-C and Birt, DF). Papilloma data from 29 weeks after DMBA; carcinoma data from 40 weeks after DMBA.

Terpenes. Another class of compounds which holds considerable
promise as inhibitors of cancer are the terpenes. D-limonene, which
is a component of citrus oils, is the terpene which probably has
been studied most extensively as an inhibitor of carcinogenesis.
Early studies reported that D-limonene, a mixture of D-limonene with
its hydroperoxide, and orange oil were similar in their ability to
prevent subcutaneous tumors induced by benzo(rst)pentaphene (DBP)
(56) but that spontaneous and DBP-induced lung adenomas were reduced
in mice fed D-limonene but not in those fed orange oil or the hydroperoxide of limonene.

Studies on DMBA-induced mammary carcinogenesis in Sprague-
Dawley rats indicated that feeding diet containing 1 or 10 g/kg D-
limonene from 1 week before DMBA until the end of the experiment
resulted in decreased incidence and latency and increased regression
of mammary tumors (57). D-limonene was most effective when administered during initiation, but some inhibition was observed in rats
fed D-limonene during progression (58).

A recent report on the Oleanane-type synthetic triterpenoids
showed impressive suppression of tumor promotion in the DMBA-initi-
ated, TPA-promoted model of skin tumorigenesis by topical application of these compounds (59). For example, application of 81 nmol
18-α and 18β-olean-12-ene-3, β23,28-triol inhibited the incidence of
tumors by 80 and 60%, respectively. Furthermore, these compounds
also inhibited the incorporation of ^{32}P into phospholipids of cells
cultured with TPA as shown in Table VII. These results suggest that
these compounds may hold particular promise for the prevention of
cancer.

Table VII. Effects of Oleanane-Type Triterpenoids on TPA-Induced
^{32}Pi incorporation into phospholipids of cultured cells

Cells	Triterpenoid	^{32}Pi incorporation (% inhibition)
C3HlDT1/2	18β-olean-12-ene-3β,23,28-triol	50
	18α-olean-12-ene-3β,23,28-triol	80
Swiss 3T3	18β-olean-12-ene-3β,23,28-triol	59
	18α-olean-12-ene-3β,23,28-triol	89

SOURCE: Reproduced with permission from ref. 59. Copyright 1988 *Cancer Research*.

Summary and Conclusions

Dietary inhibition of cancer is being approached through the use of
modifications in macronutrient intakes, micronutrient intakes and
the consumption of non-nutrient components of diets. Current recommendations on dietary prevention of cancer emphasize reduction in
the intake of fat and increased intake of fruits and vegetables.
This chapter provides an overview of the type of data being collect-
ed in support of these recommendations and the approach being taken
to further our understanding of this area. In general, dietary

modification of cancer emphasizes inhibition of the later stages of the cancer process. It is assumed that we cannot avoid conditions which will possibly result in the initiation events of cancer. These events are caused by such a wide range of agents that it will probably not be possible to devise a diet which would prevent the initiating event. Thus, dietary modification will be most useful in preventing the development of tumors from initiated cells. It is hoped that the types of dietary modifications being studied will be applicable for the prevention of cancers potentially caused by pesticides as well as by other agents.

Literature Cited

1. Diet, Nutrition and Cancer, Committee on Diet, Nutrition and Cancer, National Research Council, National Academy Press: Washington, DC, 1982.
2. American Cancer Society. CA Cancer J. Clin. 1984, 34, 121-126.
3. Birt, D. F. Amer. J. Clin. Nutr. 1987, 45, 203-209.
4. Berg, J. W. Cancer Res. 1975, 35, 3345-3350.
5. Miller, A. B.; Howe, G. R.; Jain, M.; Craib, K. J. P.; Harrison, L. Int. J. Cancer 1983, 32, 155-161.
6. Boissonneault, G. A.; Elson, C. E.; Pariza, M. W. J. Natl. Cancer Inst. 1986, 76, 335-338.
7. Birt, D. F.; Julius, A. D.; White, L. T.; Pour, P. M. Proc. Amer. Assn. Cancer Res., 1987, Abstract 608.
8. Jacobs, L. R. Proc. Soc. Exptl. Biol. Med. 1986, 183, 299-310.
9. Potter, J. D.; McMichael, A. J. J. Natl. Cancer Inst. 1986, 76, 557-569.
10. Birt, D. F. Proc. Soc. Exptl. Biol. Med. 1986, 183, 311-320.
11. Mettlin, C.; Graham, S.; Swanson, M. J. Natl. Cancer Inst. 1979, 62, 1435-1438.
12. Peto, R.; Doll, R.; Buckley, J. D.; Sporn, M. B. Nature (London) 1981, 290, 201-208.
13. Shekelle, R. B.; Liu, S.; Raynor, W. J.; Lepper, M.; Maliza, C.; Rossof, A. H.; Paul, O.; MacMillan Shryock, A.; Stamler, J. Lancet 1981, 2, 1185-1190.
14. Wald, N.; Idle, M.; Boreham, J. Lancet 1980, 2, 813-815.
15. Moon, R. C.; McCormick, D. L.; Becci, P. J.; Shealy, Y. F.; Frickel, F.; Paust, J.; Sporn, M. B. Carcinogenesis 1982, 3, 1469-1472.
16. Cohen, S. M.; Wittenberg, J. F.; Bryan, G. T. Cancer Res. 1976, 36, 2334-2339.
17. Moon, R. C.; McCormick, D. L. J. Amer. Acad. Dermatol. 1982, 6, 809-814.
18. Welsch, C. W.; DeHoog, J. V.; Moon, R. C. Carcinogenesis 1984, 5, 1301-1304.
19. Ip, C. Carcinogenesis 1981, 2, 915-918.
20. Mirvish, S. S. In Cancer 1980: Achievements, Challenges, Projects; Burchenal, J. H.; Oettgen, H. P., Eds.; Grune & Stratton: New York, 1981; Vol. 1, pp 557-587.
21. Mirvish, S. S. J. Natl. Cancer Inst. 1983, 71, 629-647.
22. Mergens, W. J. Ann. N.Y. Acad. Sci. 1982, 393, 61-69.
23. Horvath, P. M.; Ip, C. Cancer Res. 1983, 43, 5335-5341.
24. Ip, C. Carcinogenesis 1982, 3, 1453-1456.

25. Salonen, J. T.; Alftran, G. Amer. J. Epidemiol. 1984, 120, 342–349.
26. Birt, D. F.; Pelling, J. C.; Tibbels, M. G.; Schweickert, L. Nutrition and Cancer 1988, 11, 21–33.
27. O'Connor, T. P.; Roebuck, B. D.; Peterson, F.; Campbell, T. C. J. Natl. Cancer Inst. 1985, 75, 959–962.
28. Birt, D. F. In Horticulture and Human Health: Contributions of Fruits and Vegetables; Quebedeaux, B.; Bliss, F. A., Eds.; Prentice-Hall, Englewood Cliffs, N.J., 1987; pp 160–173.
29. Hu, J.; Zhang, S.; Jia, E.; Wang, Q.; Liu, S.; Liu, Y.; Wu, Y.; Cheng, Y. Int. J. Cancer 1988, 41, 331–335.
30. Stoewsand, G. S.; Anderson, J. L.; Munson, L. Cancer Lett. 1988, 39, 199–207.
31. Godlewski, C. E.; Boyd, J. N.; Sherman, W. K.; Anderson, J. L.; Stoewsand, G. S. Cancer Lett. 1985, 28, 151–157.
32. Scholar, E. M.; Wolterman, K.; Birt, D.; Bresnick, E. Nutr. Cancer, in press, 1989.
33. Birt, D. F.; Pelling, J. C.; Pour, P. M.; Tibbels, M.G.; Schweickert, L.; Bresnick, E. Carcinogenesis 1987, 8, 913–917.
34. Temple, N. J.; Basu, T. K. J. Natl. Cancer Inst. 1987, 79, 1131–1134.
35. Goeger, D. E.; Shelton, D. W.; Hendricks, J. D.; Bailey, G. S. Carcinogenesis 1986, 7, 2025–2031.
36. Bailey, G. S.; Hendricks, J. D.; Shelton, D. W.; Nixon, J. E.; Pawlowski, N. E. J. Natl. Cancer Inst. 1987, 78, 931–934.
37. Ansher, S. S.; Dolan, P.; Bueding, E. Fd. Chem. Toxic. 1986, 24, 405–415.
38. Harborne, J. B.; Mabry, T. J.; Mabry, H. The Flavonoids; Academic Press: New York, 1975; pp 1011–1014, 1033–1036.
39. Swain, T. Prog. Clin. Biol. Res. 1986, 213, 1–14.
40. Pierpoint, W. S. In Plant Flavonoids in Biology and Medicine. Biochemical, Pharmacological, and Structure-Activity Relationships; Cody, V.; Middleton, E., Jr.; Harborne, J. B., Eds.; A.R. Liss Inc.: New York, 1986; pp 125–140.
41. Brown, J. P.; Dietrich, P. S. Mutation Res. 1979, 66, 223–240.
42. Pamucku, A. M.; Yalciner, S.; Hatcher, J. F.; Bryan, G. T. Cancer Res. 1980, 40, 3468–3472.
43. Hirono, I.; Ueno, I.; Hosaka, S.; Takanashi, H.; Matsushima, T.; Sugimura, T.; Natori, S. Cancer Lett. 1981, 13, 15–21.
44. Morino, K.; Matsukura, N.; Kawachi, T.; Ohgaki, H.; Sugimura, T.; Hirono, I. Carcinogenesis 1981, 3, 93–98.
45. Saito, D.; Shirai, A.; Matsushima, T.; Sugimura, T.; Hirono, I. Terat. Carc. Mutagen 1980, 1, 213–221.
46. Hirose, M.; Fukushima, S.; Sakata, T.; Inui, M.; Ito, N. Cancer Lett. 1983, 21, 23–27.
47. Takanashi, H.; Aiso, S.; Hirono, I.; Matsushima, T.; Sugimura, T. J. Food Saf. 1983, 5, 55–60.
48. Nishino, H.; Iwashima, A.; Fujiki, H.; Sugimura, T. Gann 1984, 75, 113–116.
49. Fujiki, H.; Horiuchi, T.; Yamashita, K.; Hakii, H.; Suganuma, M.; Nishino, H.; Iwashima, A.; Hirata, Y.; Sugimura, T. In Plant Flavonoids in Biology and Medicine; Cody, V.; Middleton, E., Jr.; Marbone, J. B., Eds.; Alan R. Liss, Inc.: New York, 1986; pp 429–440.

50. Johnson, J. A.; Gould, M. N.; Tanner, M. A.; Verma, A. K.;
    Madison, W. I. Proc. Amer. Assn. Cancer Res., 1988, Abstract
    517.
51. Blumberg, P. M. CRC Crit. Rev. Toxicol. 1981, 9, 199-234.
52. Diamond, L.; O'Brien, T. H.; Baird, W. M. Adv. Cancer Res.
    1980, 32, 1-74.
53. Nakadate, T.; Yamamoto, S.; Aizu, E.; Kato, R. Gann 1984, 75,
    214-222.
54. Birt, D. F.; Walker, B.; Tibbels, M. G.; Bresnick, E.
    Carcinogenesis 1986, 7, 959-963.
55. Nishino, H.; Nishino, A.; Iwashima, A.; Tanaka, K.; Matsuura, T.
    Oncology 1984, 41, 120-123.
56. Homburger, F.; Treger, A.; Boger, E. Oncology 1971, 25, 1-10.
57. Elegbede, J. A.; Elson, C. E.; Qureshi, A. Carcinogenesis 1984,
    5, 661-664.
58. Gould, M. N.; Maltzman, T. H.; Boston, J. L.; Tanner, M. A.;
    Sattler, C. A.; Elson, C. E. Proc. Amer. Assn. Cancer Res.
    1986, 27, 131.
59. Nishino, H.; Nishino, A.; Takayasu, J.; Hasegawa, T.; Iwashima,
    A.; Hirabayashi, K.; Iwata, S.; Shibata, S. Cancer Res. 1988,
    48, 5210-5215.

RECEIVED June 28, 1989

Chapter 8

Impact of Chemical Interactions on the Development of Cancer

Harihara M. Mehendale

Department of Pharmacology and Toxicology, University of Mississippi
Medical Center, 2500 North State Street, Jackson, MS 39216–4505

Relatively few toxic chemical interactions of carcinogenic consequence involving pesticide chemicals
have been studied. Such interactions may result
in a decreased, unaltered, or increased risk of
cancer and all interactions are of interest in risk
assessment. Decreased risk of cancer may be associated with increased mortality due to potentiated
toxicity or due to the altered toxicokinetics of the
carcinogen, resulting in an overall increase in
risk. An increased risk of cancer may be evident in
the form of increased incidence or severity, and may
involve the same target organ or new target tissues. Examples of the types of interactions include co-carcinogenicity, promotion or potentiation
of cancer involving genotoxic or epigenetic mechanisms. Understanding the underlying mechanisms
is essential to developing strategies for prediction of notorious interactions as well as minimizing the risks to public health.

There is a significant and growing interest in the toxicology of
exposure to combinations of chemicals (1-15). Human and animal
populations are simultaneously or sequentially exposed to pure
single- or multiple-component chemicals such as formulations.
Even if one considers a single-component chemical, the
toxicological considerations are usually compounded by exposures
to one or more other toxic chemicals in our environment (4).
The most potent interaction of toxicological significance
reported to date involves two known animal carcinogens (2).
Whether this interaction between chlordecone and CCl_4 results in
amplified carcinogenicity has not been tested. The interaction
represents a sequential exposure to a halogenated pesticide,
chlordecone and a haloalkane, both at individually nontoxic
doses (10).

0097–6156/89/0414–0122$06.00/0

The primary objective of this chapter is to consider how other chemicals may influence carcinogenic response of pesticide chemicals. However, a broader consideration of how pesticidal compounds may affect the outcome of the response to a known carcinogen is also of significant interest. Therefore, a general discussion of how the presence of pesticidal compounds may affect the outcome of carcinogenic response regardless of the compound or agent initiating the response will be worthwhile. From the public health point of view, of primary interest are those interactions, which increase carcinogenicity. In the interest of unravelling the underlying mechanisms, interactions with pesticides or other chemicals resulting in unaltered or even decreased carcinogenic responses must also be studied.

Mechanisms of Interactions

Chemical agents enhance and inhibit chemical carcinogenesis by a variety of mechamisms (Table I). These include modification of carcinogen availability, bioactivation, reactive interactions, depletion of cytoprotective cellular nucleophiles, and elimination processes.

Table I. Mechanisms by which chemicals may alter
carcinogenic response

I. Carcinodynamic Modulation

   Altered bioavailability
   Altered bioactivation
   Depletion of protective nucleophiles

II. Physiological modulation of the target

   Inhibition or induction of DNA repair
   Inhibition or enhancement of cellular
      proliferation

Since these mechanisms involve alterations in the nature and quantity of the utimate carcinogen at the target tissue, these may be considered as carcinodynamic mechanisms. Modulation of the target of carcinogenic response may also be brought about by other chemicals. These include interference with DNA repair mechanisms, and at the cellular level, analogous effects on cellular proliferation.

   A chemical agent may alter the absorption of a cancer causing chemical at the point of entry or at the level of cellular uptake, can alter carcinogenicity. An example of altered absorption may be found in the latency period in the development of subcutaneous neoplasm by the same dose of benzo(rst) pentaphene applied in different vehicles. The latency period was 16 weeks in peanut oil, 37 weeks in lipoprotein, and 62 weeks in Ringer's solution (16). Likewise, in a study of gastric carcinogenesis by N-methyl-N-nitro-N-nitrosoguanidine (17), the incidence of gastric tumors with olive oil was 70%, while with saline it was 36%.

The carcinogenicity of genotoxic procarcinogens can be increased or decreased by modification of the enzyme systems involved in their biotransformation (18-21). Several pesticidal compounds known to induce drug metabolizing enzymes of the liver have been reported to influence the outcome of carcinogenic effect of known carcinogens (Table II).

Table II. Influence of Enzyme Inducers or Inhibitors on Carcinogenicity

Inducer	Carcinogenic Effect (Increase,↑ or Decrease,↓)	Organ	Authors
DDT	Dimethyl Benz(a) anthracene↓	Breast	Silinskas and Okey, 1975
Disulfiram	Ethylene↑ dibromide	Liver Kidney Splean	Wong et al., 1982
Toxaphene	Benzo(a)pyrene↓	Lung	Triolo et al., 1982
Carbaryl	Benzo(a)pyrene↑	Lung	Triolo et al., 1982

Dietary pretreatment with DDT decreased the incidence of mammary tumors induced by dimethyl benzanthracene (21). Carcinogenicity of ethylene dibromide has been shown (18) to be increased by exposure to disulfiram. Induction of lung benzo(a)pyrene hydroxylase by carbaryl paralleled an increase in lung tumors induced by benzo(a)pyrene (19). In the same experiments, pretreatment with toxaphene decreased the incidence of lung adenomas (19) and this correlated with the inhibition of lung benzo(a)pyrene hydroxylase.

Protection against the carcinogenicity of aminoazo dyes by riboflavin (20) is due to the action of azo reductase, which requires riboflavin as a cofactor. In the case of enzyme induction, the final outcome of carcinogenic response may depend on the balance between activation and inactivation reactions (22). The balance in the case of activation-dependent carcinogens is usually, but not always, in the direction of inactivation. The induction of the conjugating enzymes such as uridine diphosphate-glucuronyl transferase and glutathione transferase that inactivate carcinogens appears to account for the observed inhibition (23,24). The anticarcinogenic effects of phenobarbital, an inducer of drug metabolizing enzymes, against a variety of genotoxic carcinogens have been described (25-30). Although at first, such effects may appear to be paradoxical because of the well known promoter effect of phenobarbital, one may be able to explain these observations on the basis of the doses and other exposure-related factors (30). Such factors include, the dose of initiator and promotor, and the sequence of administration. For example, phenobarbital can be shown to be cocarcinogenic, anticarcinogenic or a promotor depending on these exposure related factors. Administation of phenobarbital prior to 3'-melthyl-4-(dimethyl amino) - azobenzene (3'-Me-DAB) results in the anticarcinogenic effect owing to the induction of enzymes which metabolize 3'-Me-DAB to products less likely to form DNA adducts. Simultaneous administration of 3'-Me-DAB (100 ppm) and phenobarbital (20-500 ppm) resulted in cocarcinogenic effect. When phenobarbital was given in the initiator - promotor sequence, the promotor effect was observed (30). However, such interactions are complex and may result in increased carcinogenicity in another organ as exemplified by the butylated hydroxytoluene inhibition of liver carcinogenicity and increased bladder cancer (31).

Inhibitors of enzymes involved in carcinogen metabolism can reduce the effects of activation-dependent carcinogens (13). The influence of toxaphene in decreasing the carcinogenic response of benzo(a)pyrene (Table II) is due to the inhibition of pulmonary benzo(a)pyrene hydroxylase (19). In some situations, however, inhibition of metabolism increases carcinogenic effects (32,33). Furthermore, enzyme inhibitors can produce an increase of neoplasms in secondary sites, presumably as a consequence of increased availability of the carcinogen in other organs resulting from inhibition of biotransformation in the primary organ for metabolism (18,34).

Depletion of cellular substrates involved in activation of carcinogens can result in inhibition of carcinogenicity (35-37). In these studies, it was shown that acetanilide decreased the liver carcinogenicity of 2-acetyl aminofluorine (2-AAF) by decreasing the formation of the ultimate carcinogenic metabolite, the sulfate ester of 2-AAF, mainly through depletion of the available sulfate as a result of sulfation of p-hydroxy acetanilide (36,37).

An important effect of chemicals on carcinogen metabolism is the permanent alteration of enzyme systems, known as imprinting, which is produced by exposures to chemicals during the developmental period (38,39). Rats exposed to synthetic hormones in the neonatal period have displayed altered carcinogenic responses to polycyclic hydrocarbons (40,41). The phenomenon of imprinting may also provide an explanation for cervical cancer in daughters of mothers who were administered diethyl stilbesterol during pregnancy.

Modification of reactive intermediates is another mechanism for alteration of carcinogenic response. A reactive type of carcinogen is converted to an electrophile or radical cation, molecular species capable of binding to cellular nucleophiles eventually resulting in neoplastic response. DNA is a critical target for such reactions. In addition to producing neoplasia, genotoxic carcinogens may be capable of inflicting other genotoxic or cytotoxic effects resulting in selective proliferation of initiated cells (13,42).

Nongenotoxic agents may indirectly generate reactive species such as activated oxygen (43,44). Any agent that competes with the reactive carcinogen or radical species generated in the cell might be expected to inhibit carcinogenesis. Cellular nucleophiles such as those containing sulfhydryl groups have been suggested to act in this manner (45). Glutathione, known to bind reactive metabolites of carcinogens has been reported to inhibit the liver carcinogenicity of aflatoxin B_1 (46).

DNA adducts of carcinogens can be removed by DNA repair systems and this mechanism reduces the mutagenic or carcinogenic effects. The repair of certain kinds of alkylation damage in DNA by carcinogens appears to be inducible (47,48), although whether this results in reduction of carcinogenic response is unclear. Cocarcinogens could produce their effects by interfering with DNA repair. 3-Aminobenzamide known to inhibit processes specific for DNA repair such as poly(ADP-ribose) polymerase (49) has enhanced the effect of a liver carcinogen (50).

An additive or synergistic effect of two or more carcinogens in neoplasm production is defined as syncarcinogenesis (51,52). Two variations of this type of interactions are, combination or sequential syncarcinogenesis (Table III). This type of interaction has been demonstrated with many agents affecting many organs and a variety of mechanisms have been invoked (13). An interesting syncarcinogenic effect is that produced by a genotoxic and a nongenotoxic carcinogen.

Table III.   Distinction Between Types of Multiple
             Exposures in Chemical Carcinogenesis

Process	Operational Characteristics	Mechanistic Differences
Combination syncarcinogenesis	Two carcinogens acting together	Both carcinogens genotoxic
Cocarcinogenesis	Enhancer acting before or together with carcinogen; or when carcinogen effects still persist	Enhancer facilities neoplastic; conversion nongenotoxic and noncarcinogenic
Sequential syncarcinogenesis	Two carcinogens acting in sequence; sequence reversible	Both carcinogens genotoxic
Promotion	Enhancer acting after effects of carcinogens have been completed	Enhancer facilitates neoplastic development; enhancer may also be a carcinogen, but is nongenotoxic

Adapted from reference (13) with permission.

The nongenotoxic liver carcinogen, clofibrate, which is a peroxisome proliferator, enhances the hepatocarcinogenicity of previously administered diethylnitrosamine (53,54). Pesticidal compounds of the phenoxy acetic acid class should undergo closer scrutiny for this type of interaction with genotoxic carcinogens. Similarly, the nongenotoxic liver carcinogen methapyrilene, an antihistamine known to induce mitochondrial proliferation, enhances the liver carcinogenicity of 2-AAF (55,56). Thus, indirect genotoxic effect postulated to be produced by these types of agents can be amplified as enhanced DNA damage produced by genotoxic carcinogens (57,58).

The progeny of the neoplastic cells can remain latent for long intervals (59,60), indicating the control of proliferative activity by host homeostatic mechanisms. Hence, agents capable of enhancing differentiation of neoplastic cells or facilitation of their control by tissue factors, could inhibit neoplastic development. Vitamin A and related retinoids appear to inhibit carcinogenesis through effects on cell differentiation. Other factors known to modify neoplastic proliferation are protease inhibitors (61), nonsteroidal anti-inflammatory agents such as indomethacin (62), and the immune system (63). Immunosuppressants such as azathioprine have increased the development of lymphomas and leukemias (64-66). Neoplasm promotion was originally defined conceptually as the encouragement of dormant neoplastic cells to develop into growing tumors (67) and ideally demonstrated by a promotor substance administered after the carcinogen (68). The phenomenon was first established in skin carcinogenesis (69-71); the most extensively studied promotors are phorbol esters such as phorbol myristate acetate (tetradecanoyl phorbol acetate; TPA)(68,72,73). A variety of promotors for organs such as breast, colon, liver, and bladder have been described (74,75). In addition to TPA, other promotors are, an anti-seizure drug phenobarbital, chlorinated hydrocarbons such as DDT, chlordane, heptachlor, polychlorinated biphenyls (PCBs), TCDD, the artificial sweetener saccharin, and antioxidant butylated hydroxy toluene (BHT). A variety of cellular effects have been demonstrated for promoting agents (76,77). Table IV presents some of these possible effects (13).

Table IV. Mechanisms of Promotion of Carcinogenicity

Enhanced expression of neoplasm
Inhibited differentiation
Stimulated cellular proliferation
Cytotoxic effects
Hormonal effects
Cell membrane effects
Induction of proteases
Inhibition of intercellular Communication
Immunosuppression

Adapted from reference (13) with permission.

Recently, a hypothesis based on inhibition of cell-to-cell communication has received considerable attention. According to this concept, neoplastic cells are normally restrained by interactions with normal cells and disruption of these exchanges would thereby release dormant neoplastic cells from tissue constraints allowing them to proliferate according to the altered genome (78,79). In support of this hypothesis, a variety of promotors have now been shown to produce inhibition of molecular transfer in culture (78,80-85).

It is now established that some agents with promoting activity will increase neoplasia in the absence of initiation (86,87). Promotors that enhance liver carcinogenesis are of particular interest because they include a number of pharmaceuticals, food additives, and pesticides (88-90). Exogenous hormones and other hormonally active substances are also important in this regard (91,92).

Progression is often referred to in a restricted sense to denote the change of a neoplasm from benign to malignant or from a low grade to high grade malignancy (93). Neoplastic progression defined as the stepwise development of a neoplasm through qualitatively different stages (86) includes initiation and development of neoplasms. Proposed mechanisms for the evolution of new cell types within a neoplasm include infidelity of DNA polymerases (94) and hybridization of normal and neoplastic cells (95). Since these events might be subjected to interference by other chemicals, this provides yet another mechanism for chemical modulation of the ultimate carcinogenic response.

Examples of Pesticides Affecting Carcinogenic Responses

Table V lists the classes of pesticides of which some members are known to modify the induction of cancer. Halogenated pesticides are by far the most widely studied in this regard. Examples from insecticides, fungicides, herbicides and other types of agrochemicals fall into these broad chemical classes.

Table V.   Chemical classes of Pesticides Involving
Carcinogenic Interactions

Organochlorines
Carbamates
Organophosphates
Metals
Others

Some important representative carcinogens involving carcinogenic interactions with a number of pesticides and other chemicals are listed in Table VI.

Table VI.   Pesticides and 'Other' Chemicals Involved in
Carcinogenic Interactions

DDT	Diethylnitrosamine (DENA)
α-Hexachlorocyclohexane	2-Acetylaminofluorine (2-AAF)
Dieldrin, Heptachlor,	Benzo(a)pyrene (BaP)
Chlordane, Toxaphene	
3-Me-4(dimethylamino)azobenzene	Ethylnitrosourea
7,12,-dimethylbenz(a)athracene	Malathion
Copper, Mercury	Nitrite
Others	Virus

Among the organochlorine pesticides, chlorinated insecticides
such as DDT, dieldrin, chlordane, α-hexachlorocyclohexane (a
component of BHC), lindane, provide the most widely studied
examples. Relatively few studies are available on the
carcinogenic interactions of carbamates and organophosphates.
Carbaryl and malathion are examples of carbamates and
organophosphates, respectively, from the insecticide field.
Organometallics containing copper and mercury represent metallo
fungicide class of pesticides, known to be involved in
carcinogenic interactions.

DDT and its metabolites (DDE,DDD) have been shown to be
tumorigenic in the liver individually (96-100) and in
combinations (96) in a number of strains of mice and rats (101).
DDT has also been shown to promote tumorigenic responses of 2-
AAF (102) and diethylnitrosamine (103,104) in rats and
dimethylnitrosamine (105) in mice. In rats, previously fed a
diet containing 2-AAF, both phenobarbital and DDT were shown to
promote tumorigenesis (102). Phenobarbital and DDT both
increased early tumor incidence rate and maintained an increment
in tumor incidence rate over other groups treated with diphenyl
hydantoin and amobarbital. In addition, although the spectrum
of tumor types observed ranged from highly differentiated to
poorly differentiated in all treatment groups, DDT and
phenobarbital selectively increased the incidence of highly
differentiated tumors (102). Others (103) have also shown DDT
to promote diethylnitrosamine-induced hepatocellular carcinoma
in rats. In the same study, the incidence of lung tumors was
decreased by DDT. Species differences exist in the promotor
capability of DDT. Diethylnitrosamine-induced carcinogenesis
was not enhanced by subsequent administration of phenobarbital
or DDT in hamsters (106).

Evidence has accumulated in support of the hypothesis
(78,79) that DDT inhibits metabolic cooperation (84,107) thereby
causing a loss of control on cell proliferation. DDT was shown
to inhibit cell-to-cell communication between primary cultures
of hepatocytes and an established adult rat liver epithelial
cell, 6-thioguanine resistant strain (84). Recently, DDT was
compared with a phorbol ester and quercetin in the same system
and the phrobol ester and DDT were found to be synergistic
indicating that they may act by different pathways (107). The
well established promotor action of DDT appears to be consistent
with its capacity to inhibit cell-to-cell communication.

DDT pretreatment has been reported to decrease the tumorigenic response of 7,12-dimethylbenz(a)anthracene (21) and 3-methyl-4-(dimethylamino) azobenzene (30). This effect appears to be mediated by the enhancement of the enzyme systems responsible for the inactivation of the carcinogens. Other inducers of drug metabolizing enzymes also share this property (13,30).

An interesting study involving the pesticide ethylene dibromide and disulfiram provides a good example of the type of carcinogenic interaction (18), where aldehyde dehydrogenase, an enzyme involved in the metabolism of ethylene dibromide may be inhibited by disulfiram. Ethylene dibromide is a soil fumigant nematocide, and possible adverse effects after occupational exposure prompted this investigation in which disulfiram was employed. Male and female rats were exposed to ethylene dibromide by inhalation (20 ppm) up to 18 months and disulfiram was included in the diet at 0.05%. Acetaldehyde dehydrogenase plays a role in the metabolism of ethylene dibromide and disulfiram is a known inhibitor of this enzyme. The histopathological findings are summarized in Table VII.

Table VII. Major Histopathological Findings in Rats Exposed to Ethylene Dibromide (EDB) Alone or in Combination with Disulfiram(DS)

| | EDB | | EDB+DS | |
	Male	Female	Male	Female
No. of Animals Examined	46	48	48	45
Liver hepatocellular tumors	2	3	36*	32*
Mesentary or omentum Hemangiosarcoma	0	0	11*	8*
Kidney adenoma and adenocarcinoma	3	1	17*	7*
Thyroid follicular epithelial adenoma	3	1	18*	18*
Mammary all tumors	---	25	---	13*
Lung all tumors	3	0	9*	2
No. of Rats with Tumors	25	29	45*	45*
No. of Rats with Multiple Tumors	10	8	37*	32*

Adapted from reference (18) with permission

Combined exposure resulted in decreased latency of tumori-genesis, greater incidence and tumors were found in most tissues in both the sexes. Mortality was also increased significantly and death occurred much earlier (18). The interaction not only increased the hepatocellular carcinoma, but presumably the inhibition of ethylene dibromide metabolism by disulfiram and possibly increased distribution to other tissues, resulted in increased tumor incidence at other sites as well (Table VII).

α-Hexachlorocyclohexane causes hepatocellular carcinoma (99,108,109) and has been reported to promote the development of liver tumors from precarcinogenic tumors (110,111). The promoting action of this pesticide was also demonstrated with diethylnitrosamine-induced hepatocellular carcinoma (112). Schulte-Hermann and Parzell (111) have suggested that hexachlorocyclohexane is not a tumor initiating carcinogen and that its tumorigenic capacity may simply be a reflection of the promotion of "spontaneous" lesions. It exhibits specificity for liver tumor promotion and this has come to light due to its failure to promote dimethylbenz(a)anthracene-initiated skin tumors (112).

Dieldrin, heptachlor and chlordane are cyclodiene halogenated hydrocarbons, and have been shown to be tumorigenic (105,113,114). In similarity to hexachlorocyclohexane, dieldrin has also been considered to exacerbate or facilitate the expression of a preexisting oncogenic factor possibly of viral origin (113,114). A recent report (115) indicates that dieldrin increases the susceptibility of peritoneal macrophages to mouse hepatitis virus, supporting the suggestion by Tennekes et al. (113) that the genetically linked carcinogenic factor might be of viral origin.

Hexachlorobenzene has been reported to be carcinogenic in hamsters (116). In mice, this compound enhances the carcinogenic response of polychlorinated terphenyls (117). Another fungicide, thiabendazole is also reported to enhance the bladder carcinogenic effect of sodium o-phenylphenate (118).

Carbaryl, a carbamate insecticide has not been associated with carcinogenicity. However, the compound has been reported to increase the cell associated Varicella-Zoster virus (119). Virus replication occurred two ways. More virus was taken up by the cells and the interferon response of the treated cells was decreased. A dose-response study with carbaryl showed a decrease in viral enhancement in cells treated with decreasing carbaryl concentrations. Mice pretreated with carbaryl suffer subsequently from greater mortality rates when challenged with nonlethal doses of encephalomyocarditis virus (120). The organophosphate, malathion was shown to increase myelogenous leukemia induced by repeated administration of 7, 12-dimethylbenz(a)anthracene (21). Animals given malathion had a higher mammary tumor incidence, shortened latency period, more tumors per rat, and more actively growing tumors (21). This effect is presumably due to the inhibition of the carcinogen inactivation resulting in greater tissue distribution of the carcinogen.

Organometallics have been reported to increase carcinogenic responses of known carcinogens in several studies. Mercury chloride has been shown to increase the transplacentally induced tumors by ethyl nitrosourea (121). The latency period

for neurogenic tumors was shortened. Ependymomas was found only
in progeny from rats given methyl mercury
plus sodium nitrite and ethyl urea, and the incidences of
schwannomas of the cranial nerves were consistently higher in
progeny from these rats.
     Piperonyl butoxide, a prototype of methylene dioxyphenyl
synergists has been extensively tested for carcinogenicity
(122,123) and the findings appear to be negative.
     A number of pesticides (124-127) have been examined in a
variety of *in vitro* systems either alone or in combinations to
examine their mutagenic potential. In a detailed study, 65
pesticides were classified into chemical structure profiles in
accordance with the findings of batteries of *in vitro*
mutagenesis assays. Three major classes were: a, those which
cause DNA damage, gene mutation, and chromosomal effects; b,
those causing mutation in mammalian cells and chromosomal
effects; c, those causing DNA damage in eukaryotes. These
studies will not be discussed any further since these do not
directly predict carcinogenic responses, let alone predicting
the qualitative and quantitative aspects of how those might be
altered. A number of other compounds including herbicidal and
insecticidal carbamates, have been examined for the possibility
of *in vivo* nitrosation after combined administration to animals
(128-129). Although some evidence is available to suggest that
in vivo nitrosation may occur giving rise to nitroso derivatives
(128), the implication of these products for carcinogenic
responses is far from clear.
     Pesticidal compounds may interact with carcinogens in a
variety of ways. These mechanisms are listed in Table VIII.

        Table VIII. Proximal Mechanisms Involved in Carcinogenic
                        Interactions of Pesticides

Promotion:
        Initiated preneoplastic lesion is promoted by pesticides.
        Activation of preexisting factors; Oncongenes
Initiation:
        Formation of product(s) capable of initiation and
                promotion.

Amplification of Haloalkane Toxicity by Chlordecone.

An example of a pesticide shown to be carcinogenic in animal
studies (131) potentiating the toxicity of known animal
carcinogens such as CCl_4 (132) and $CHCl_3$ (133) has come to light
(2,10). This interaction is highly unusual and the underlying
mechanism is of significant interest. An ordinarily nontoxic or
subtoxic dose of haloalkane becomes highly hepatotoxic (134,135)
by prior exposure to nontoxic levels of chlordecone (134,135).
The remarkably amplified lethality (Table IX; 2,10, 137-141) is
unprecedented. The highly unusual aspect of the interaction is
that mirex, a closely related structural analog does not
potentiate haloalkane hepatotoxicity with nearly the propensity

displayed by chlordecone (135-137,142). The mechanism underlying this remarkable amplification of haloalkane toxicity has eluded a number of investigations (143-149).

Table IX.   Amplification of Lethal Effects of Several Haloalkanes by Subtoxic Dietary Levels of Chlordecone

Dietary Pretreatment	Haloalkane	48hr LD_{50} ml/kg	95% Confidence Limits	Decrease in LD_{50}
Female Rats				
Control	CCl_4	1.25	0.85-1.84	—
Chlordecone (10ppm)	CCl_4	0.048*	0.03-0.07	26-fold
Male Rats				
Control	CCl_4	2.8	1.5-5.2	—
Chlordecone (10ppm)	CCl_4	0.042*	too low to calculate	67-fold
Phenobarbital (225ppm)	CCl_4	1.7	1.2-2.3	1.6-fold
Control	$BrCCl_3$	0.119	0.056-0.250	—
Chlordecone (10ppm)	$BrCCl_3$	0.027*	0.018-0.038	4.5-fold
Control	$CHCl_3$	0.067	0.65-0.70	—
Chlordecone (10ppm)	$CHCl_3$	0.16*	0.12-0.23	4.2-fold
Mirex (10ppm)	$CHCl_3$	0.70	0.46-1.22	No change
Phenobarbital (225ppm)	$CHCl_3$	0.70	0.41-1.18	No Change

Mechanisms such as induction of microsomal cytochrome P-450 by chlordecone, whereby enhanced metabolic bioactivation of CCl_4 could be invoked, and enhanced lipid peroxidation are inadequate to explain the remarkably powerful potentiation of toxicity and lethality. Treatment with phenobarbital, which results in a 3-fold increase in the metabolism of CCl_4 is associated with a 40-fold lesser potentiation of toxicity than prior exposure to chlordecone, which causes a lesser increase in the metabolism of CCl_4 (144,145). The extent of lipid peroxidation in the liver tissue in the two instances is very similar.

    The overall toxicity manifested in animals receiving the chlordecone + CCl_4 combination is that of a massive dose of CCl_4, even though an ordinarily nontoxic dose of CCl_4 is administered. A novel insight for the underlying mechanism was found in recent studies (2,10). Time-course histomorphometric studies in which liver tissue was examined 1 to 36 hours after CCl_4 administration revealed an answer to the question of "why is normally a low dose of CCl_4 not toxic?" While the animals receiving normally nontoxic dose of CCl_4 alone show limited hepatocellular necrosis by 6 hours, proceeding to greater injury after 12 hours, recovery phase ensues as revealed by greatly increased mitotic figures in the liver tissue. Hepatocellular regeneration continues during the progressive phase of limited injury until complete recovery is achieved well before 36 hours. During this recovery period, dead cells are replaced by newly divided cells and this is also accompanied by a restoration of normal levels of glycogen and lipid. Therefore, the primary reason for the non-toxic nature of the low dose of CCl_4 is the ability of the liver tissue to respond by hepatocellular regeneration followed by tissue repair. Stimulated hepatocellular regeneration within 6 hours after the administration of CCl_4 seems to accomplish two important objectives: first, replacement of dead cells with new healthy cells; second, the newly divided cells are more resistant to the toxic action of CCl_4. Both of these objectives are achieved by a single biological event of stimulated hepatocellular regeneration, constituting the recovery phase initiated early after the administration of CCl_4 (2,10).

Such regeneration and hepatic tissue repair processes are totally suppressed in animals exposed to chlordecone prior to the administration of a same dose of CCl_4 (150). Thus, the arrested hepatocellular regeneration and tissue repair play a key role in the uncontrollable progression and consequently, a remarkable potentiation of liver injury. These findings allow one to propose a novel hypothesis for the mechanism of chlordecone amplification of halomethane toxicity and lethality. While limited injury is initiated by the low dose of CCl_4 by bioactivation followed by lipid peroxidation, this normally recoverable injury is allowed to progress due to arrested hepatocellular regeneration and tissue repair processes. Recent studies designed to test this hypothesis have provided supporting evidence. Hepatocellular regeneration stimulated by partial hepatectomy was unaffected by 10 ppm dietary chlordecone, while these animals were protected from the hepatotoxic and lethal actions of CCl_4 if administered at the time of maximal hepatocellular regeneration (150). The protection was abolished when CCl_4 was administered upon cessation of hepatocellular regeneration (150-153). The slightly increased bioactivation of CCl_4 by prior exposure to chlordecone results in slightly increased initial injury. However, the primary reason for the highly amplified liver injury, which culminates in greatly enhanced lethality, is the progressive and irreversible phase of hepatotoxicity. This progression of hepatotoxicity is due to the suppression of the protective response of the liver tissue, namely, early stimulation of hepatocellular regeneration. Since tissue repair and renovation cannot take place, injury progresses.

Furthermore, the ability of the tissue to overcome the initial toxicity is also mitigated by the lack of newly dividing and relatively resistant cells.

While the mechanistic inquiry into this most unusual toxic interaction between two carcinogenic compounds resulting in most severe toxic response of biological significance continues, the question of whether the carcinogenic response of the haloalkanes is altered by exposure to chlordecone has not been investigated. Such an inquiry in future studies would be of interest since exposure to extremely low levels of the pesticide and the haloalkanes would be required to maintain survival of the animals long enough to permit detection of any carcinogenic responses.

In conclusion, it can be stated that a number of examples of pesticides acting as promotors are available. Some progress has also been made in understanding the mechanisms involved in the promotion of neoplasms. There is compelling evidence to suggest that chlorinated hydrocarbons such as dieldrin and α-hexachlorocyclohexane might represent those agents, which activate preexisting carcinogenic factors. Whether these compounds activate genetic factors such as oncogenes or some viral activity, remain to be investigated. Carbamates represent one class of pesticides capable of forming products which have the potential to initiate carcinogenic activity. Whether this in fact occurs *in vivo* is not entirely clear at this moment. Unprecedented toxicological interaction leading to acute hepatotoxic response and lethality has been reported with a chlorinated pesticide, chlordecone and several haloalkanes at ordinarily, individually subtoxic or nontoxic doses. Whether such an interaction also leads to enhanced carcinogenic response remains to be investigated. Future advances in our understanding in these and related areas will undoubtedly enable us to evaluate potential interactions of carcinogenic risk involving pesticidal chemicals.

Acknowledgment

The author's efforts were supported by the Burroughs Wellcome Fund and a grant from the Air Force Office of Scientific Research, AFOSR-88-0009. The author is the recipient of the 1988 Burroughs Wellcome Toxicology Scholar Award.

Literature Cited

1. Yang, R.S.H. In Pesticides: Minimizing the risks; ACS Symposium Series, No. 336, Ragsdale, N.N., Kuhr, R. J. eds, American Chemical Society, Washington, D.C., 1987; Chap 3, 20-36.
2. Mehendale, H. M.; Rev. Biochem. Toxicol. 1989, 10, 91-138.
3. Sandhu, S. S., De Marini, D. M., Mass, M. J., Moore, M. M., Manford, J. L., Ed; Short-term Bioassays in the Analysis of Complex Environmental Mixtures V.; Environmental Science Series, Plenum Press, New York, NY, 1987; Volume 36, pp 409.

4. Bingham, E.; Morris, S. Fundam. Appl. Toxicol. 1988, 10,
    549-552.
5. Scala, R. A. Fundam. Appl. Toxicol. 1988, 10, 553-562.
6. Plaa, G. L. Fundam. Appl. Toxicol. 1988, 10, 563-570.
7. Lewtas, J. Fundam. Appl. Toxicol. 1988, 10, 571-589.
8. Carter, W. H., Jr.; Carchman, R. A. Fundam. Appl. Toxicol.
    1988, 10, 590-595.
9. Berndt, W. O. Fundam. Appl. Toxicol. 1984, 4, 293-294.
10. Mehendale, H. M. Fundam. Appl. Toxicol. 1984, 4, 295-308.
11. Ackerman, D. M.; Hook, J. B. Fundam. Appl. Toxicol. 1984,
    4, 390-314.
12. Cohen, S. D. Fundam. Appl. Toxicol. 1984, 4, 315-324.
13. Williams, G. M. Fundam. Appl. Toxicol. 1984, 4, 325-344.
14. Sharma, R. P. Fundam. Appl. Toxicol. 1984, 4, 345-351.
15. Ritter, E. J. Fundam. Appl. Toxicol. 1984, 4, 352-359.
16. Homburger, F. and Tregier, A. In Progress in Experimental
    Tumor Research; Homburger F. Ed.; Karger, New York, NY
    1969; Vol. 11, 86-99.
17. Hirono, I. and Shibuyo, C. In Topics of Chemicals
    Carcinogenesis; Nakahara, W. Takayama, S., Sugimura, T.,
    Odashima, S., Ed.; Univ. of Tokyo Press, Japan, 1972;
    121-131.
18. Wong, L.C.K.; Winston, J. M.; Hong, C.B., Plotnick, H.
    Toxicol. Appl. Pharmacol. 1982, 63, 155-165.
19. Triolo, A. J.; Lang, W. R.; Coon, J. M.; Lindstrom, D.;
    Herr, D. L. J. Toxicol. Env. Health 1982, 9, 637-649.
20. Kensler, C. J.; Sugimura, K.; Young, N. F.; Halter C. R.;
    Rhoades, C. P. Science 1941, 93, 308-310.
21. Silinskas, K. C.; Okey, A. B. J. Natl. Cancer Inst. 1975,
    55, 653-657.
22. Weisburger J. H. and Williams, G. M. In Cancer Medicine;
    Holland, J. F., Frei, F. Ed., Lea & Fabiger, Philadelphia
    1982; 2nd ed.; 42-95.
23. Hesse, S., Jernstrom, G., Martinez, M., Moldeus, P.,
    Christodoulides, L. and Ketterer, B. Carcinogenesis 1982,
    3, 757-760.
24. Sparnins, V. L., Venegas, P. L., and Wattenburg, L. W. J.
    Natl. Cancer Inst. 1982, 68, 493-496.
25. Ishidate, M.; Watanabe, M.; Odasima, S. Gann Monogr.
    Cancer Res. 1967, 58, 267-281.
26. Peraino, C.; Fry, R. J. M.; Staffeldt, E. Cancer Res.
    1971, 31, 1506-1512.
27. McLean, A. E. M.; Marshall, A. K. Br. J. Exp. Pathol.
    1971, 52, 322-329.
28. Makiura, S.; Aoe, H.; Sugimura, S.; Haro, K.; Arai, M.;
    Ito, N. J. Natl. Cancer Inst. 1974, 53, 1253-1257.
29. Maeura, Y.; Weisburger, J. H.; Williams, G. M. Cancer Res.
    1984, 44, 1604-1610.
30. Kitagawa, T. Toxicol. Pathol. 1986, 14, 309-314.
31. Williams, G. M.; Maeura, Y.; Weisburger, J. H. Cancer
    Lett. 1983, 19, 55-60.
32. Kotin, P.; Falk, H. L.; Miller A. J. Natl. Cancer Inst.
    1962, 28, 725-745.

33. Argus, M. F.; Hoch-Ligeti, C.; Arcos, J. C.; Conney, A. H. J. Natl. Cancer Inst. 1978, 61, 442-449.
34. Fiala, E. S.; Weisburger, J. H.; Katayama, S.; Chandrasekharan, V.; Williams, G. M. Carcincogenesis 1981, 2, 965-969.
35. Puron, R., and Firminger, H. L. J. Natl. Cancer Inst. 1965, 35, 29-37.
36. Yamamoto, R. S.; Glass, R. M.; Frankel, H. H.; Weisburger, E. K.; Weisburger, J. H. Toxicol. Appl. Pharmacol. 1972, 13, 108-117.
37. Weisburger, J. H.; Yamamoto, R. S.; Williams, G. M.; Grantham, P. H.; Matsushima, T.; Weisburger E. K. Cancer Res. 1972, 32, 491-500.
38. Einarson, K.; Gustofson, J.; Steinberg, A. J. Biol. Chem. 1973, 248, 4987-4997.
39. Lucier, G. W. Environ. Health Persp. 1976, 29, 7-16.
40. Weisburger, E. K.; Yamamoto, R. S.; Glass, R. M.; Grantham, P.H.; Weisburger, Endocrinology J. H. 1968, 82, 685-692.
41. Rustia, M.; Shubik, P. Cancer Res. 1979, 39, 4636-4644.
42. Farber, E. In Molecular and Cellular Aspects of Carcinogen Screening Tests; Montesano, R., Barsch, H., Tomatis L., Davis, W. Ed., Academic Press, New York, NY 1979; 143-151.
43. Mason, R. P.; Chignell, C. P. Pharmacol. Rev. 1982, 33, 1989-212.
44. Moody, C. S.; Hassan, H. M. Proc. Natl. Acad. Sci. 1982; 79, 2855-2859.
45. Miller, E. C.; Miller, J. A. In Environment and Cancer, Twenty-fourth Annual Symposium on Fundamental Cancer. The University of Texas, M. D. Anderson Hospital and Tumor Institute, William & Wilkins, Maryland 1972; 5-39.
46. Novi, A. M. Science 1981, 212, 541-542.
47. Montesano, R.; Bresil, H.; Margison, G. P. Cancer Res. 1979, 39, 1798-1802.
48. Swenberg, J. A.; Bendell, M.A. Billings, K. C.; Umbenhauer, D. R.; and Pegg, A. E. Proc. Natl. Acad. Sci. 1982, 79, 5499-5502.
49. Purnell, M. R. Whish, W. J. D. Biochem. J. 1980, 185, 775-777.
50. Takahashi, S.; Ohnishi, T.; Denda, A.; Konishi, Y. Chem-Biol. Interact. 1982, 39, 363-368.
51. Nakahara, W. In Chemical Tumor Problems, Nakahara, W. Ed.; Japanese Society for Promotion of Science, Tokyo 1970; 287-330.
52. Schmahl, D. Arch. Toxicol. (Suppl.) 1980, 49, 29-40.
53. Reddy, J. K.; Rao, M. S. Br. J. Cancer 1978, 38, 537-543.
54. Mochizuki, Y.; Furukawa, K.; Sawada, N. Carcinogenesis 1982, 3, 1027-1029.
55. Furuya, K.; Williams, G. M. Toxicol. Appl. Pharmacol. 1984, 74, 63-67.
56. Furuya, K.; Mori, H.; Williams G. M. Toxicol. Appl. Pharmacol. 1983, 70, 49-56.

57. Reddy, B. S.; Cohen, L. A.; McCoy, G. D.; Hill, P.; Weisburger, J. H.; Wynder E. L. Adv. Cancer Res. 1980, 32, 237-345.
58. Rexnik-Schuller, H. M.; Lijinsky, W. Arch. Toxicol. 1981, 49, 79-83.
59. Wheelock, E. F.; Weinhold, K. L.; Levich, J. Adv. Cancer Res. 1981, 34, 107-140.
60. Sporn, M. B. In Carcinogenesis: Modifiers of Carcinogenesis, Slaga, T. J. Ed., Raven Press, New York, 1980; vol. 5, 99-109.
61. Rossman, T. G.; Trol, W. In Carcinogenesis: Modifiers of Chemical Carcinogenesis, Slaga, T. J. Ed., Raven Press, New York, 1980; Vol. 5, 73-81.
62. Kudo, T., Narisawa, T. and Abo, S. Gann 1980, 71, 260-240.
63. Pollard, M.; Luckett, P. H. Cancer Treat. Rep. 1980, 64, 1323-1327.
64. Weisburger, E. K. Cancer 1977, 40, 1935-1951.
65. Mitrou, P. S. Fisher, M.; Mitrou, G.; Rottger, P.; and Holtz, G. Arzneim. Forshch. Drug Res. 1979, 29, 483-488.
66. Imamura, N.; Nakano, M.; Kawase, A.; Kawamura, Y.; Yokoro, K. Gann 1973, 64, 493-498.
67. Berenblum, I. In Carcinogenesis as a Biological Problem, Neuberger, A., Tatum, E. L. Ed., Amer. Elsevier, New York, 1974.
68. Boutwell, R. K. CRC Crit. Rev. Toxicol. 1974, 2, 419-433.
69. Twort, S. M.; Twort, C. C. Amer. J. Cancer 1939; 35, 80-85.
70. Rous, P., Kidd, J. G. J. Exp. Med. 1941, 73, 365-390.
71. Berenblum, I. Cancer Res. 1941, 1, 807-814.
72. Hecker, E. In Carcinogenesis: A Comprehensive Survey, Slaga, T. J., Sivak, A., Boutwell, R. K. Ed., Raven Press, New York 1978, Vol. 2, 11-48.
73. Diamond, L.; O'Brien, T. G.; Baird, W. M. In Advances in Cancer Research, Klein, G., Weinhouse, S. Ed.; Academic Press, New York 1980; Vol 32, 1-74.
74. Slaga, T. J.; Sivak, A.; Boutwell, R. K.; Ed., In Carcinogenesis: A Comprehensive Survey, Raven Press, New York, 1978; Vol. 2, 11-48.
75. Hecker, E.; Fuseng, N. E.; Kunz, W.; Marks, F.; Thielmann, H. W., Ed.; Cocarcinogenesis and Biological Effects of Tumor Promotors, Raven Press, New York, 1982.
76. Weinstein, I. B.; Yamasaki, H.; Wigler, M.; Lee, L. S.; Fisher, P. B., Jeffrey, A. M., and Grunberger, D. In Carcinogenesis, Identification and Mechanisms of Action, Griffen, A. C., Shaw, C. R. Ed., Raven Press, New York, 1979; 399-418.
77. Colburn, N. H. In Carcinogenesis, Modifiers of Chemical Carcinogenesis, Slaga, T. J. Ed.; Raven Press, New York, 1980; 36-56.
78. Trosko, J. E.; Yotti, L. P.; Dawson, B.; and Chang, C. C. In Short Term Tests for Chemical Carcinogenesis, Stich, H. F., San, R. H. C. Ed., Springer-Verlag, New York, 1981; 420-427.
79. Williams, G. M. Food Cosm. Toxicol. 1981, 19, 577-583.

80. Murray, A. W.; Fitzgerald, D. J. Biochem. Biophys. Res. Commun. 1979; 91, 385-401.
81. Yotti, L. P.; Chang, C. C.; Trosco, J. E. Science 1979, 220, 1089-1091.
82. Umeda, M.; Noda, K.; Ono, T. Gann 1980, 71, 614-620.
83. Williams, G. M. Ann. N. Y. Acad. Sci. 1980, 349, 273-282.
84. Williams, G. M.; Telang, S.; Tong, C. Cancer Lett. 1981, 11, 339-344.
85. Telang, S.; Tong, C.; Williams, G. M. Carcinogenesis 1982, 3, 1175-1178.
86. Foulds, L. Ed., Neoplastic Development, Academic Press, New York, 1969.
87. Williams, G. M.; Kayayama, S.; Ohmori, T. Carcinogenesis 1981, 2, 1111-1117.
88. Peraino, C.; Fry, R. J. M.; Grube, D. D. In Carcinogenesis, A Comprehensive Survey, Slaga, T. J., Sivak, A., Boutwell, R. K. Ed., Raven Press, New York, 1978, Vol. 2, 421-432.
89. Pitot, H. C.; Goldsworthy, T.; Morgan, S.; Sirica, A. E.; Weeks, J. In Carcinogenesis, A Comprehensive Survey, Slaga, T. J., Sivak, A., Boutwell, R. K. Eds, Raven Press, New York, 1982; Vol 7, 85-98.
90. Williams, G. M. Environ. Health Persp. 1983; 50, 177-183.
91. Williams, G. M. Lab. Inv. 1982, 46, 352-354.
92. Yager, J. D. Jr. Environ. Health Persp. 1983, 50, 109-112.
93. Pitot, H. C. Fundamentals of Oncology, Decker, New York, NY, 1978.
94. Springgate, C. F.; Loeb, S. A. Proc. Natl. Acad. Sci. 1973, 70, 245-249.
95. Goldenberg, D. M.; Pavia, R. A.; Tsao, M. C. Nature (London) 1974, 250, 649-651.
96. Tomatis, L.; Turuslov, V.; Charles, R. T.; Boicchi, M.; J. Natl. Cancer Inst. 1974, 52, 883-891.
97. Innes, J. R.; Ulland, B. M.; Valerio, M. G.; et al.; J. Natl. Cancer Inst. 1969, 42, 1101-1114.
98. Walker, A. I.; Thorpe, E.; Stevenson, D. E. Food Cosm. Toxicol. 1973, 11, 415-432.
99. Thorpe, E.; Walker, A. I. Food Cosm. Toxicol. 1973, 11, 415-432.
100. Turuslov, V.S.; Day, N.E.; Tomatis, L. et al.; J. Natl. Cancer Inst. 1973, 51, 983-997.
101. Barbieri, O.; Rossi, L.; Cabral, J. R. P.; Leonardo, S. Cancer Lett. 1983, 20, 223-229.
102. Peraino, C.; Fry, M. R. J.; Staffeldt, E.; Christopher, J. P. Cancer Res. 1975, 35, 2884-2890.
103. Shivapurkar, N.; Hoover, K. L.; Poirer, L. A. Carcinogenesis 1986, 7, 547-550.
104. Nishizumi, M. Gann 1979, 70, 835-837.
105. Williams, G. M.; Numoto, S. Carcinogenesis 1984, 5, 1689-1696.
106. Tanaka, T.; Mori, H.; Williams, G. M. Carcinogenesis 1987, 9, 1171-1178.
107. Warngard, L.; Flodstorm, S.; Ljungquist, S.; Ahlborg, U. G. Carcinogenesis 1987, 8, 1201-1205.

108. Nagasaki, H.; Marugami, M.; Tomi, S.; Mega, T.; Ito, N.
     Gann 1971, 62, 431-437.
109. Nigam, S. K.; Bhat, D. K.; Karnik, A. B.; Thakore, K. N.;
     Arvind Babu, K.; Lakkad, B. C.; Kashyap, S. K.;
     Chatterjee, S. K. J. Cancer Res. Clin. Oncol. 1981, 99,
     143-152.
110. Schulte-Hermann, R.; Ohde, G.; Schuppler, J.; Timmermann-
     Troisiener, I. Cancer Res. 1981, 41, 2556-2562.
111. Schulte-Hermann, R.; Parzell, W. Cancer Res. 1981, 41,
     4140-4146.
112. Munir, K. M.; Rao, K. V.; Bhide, S. V. Carcinogenesis
     1984, 5, 479-481.
113. Tennekes, H. A.; Wright A. S.; Dix, K. M.; Koeman, J. H.,
     Cancer Res. 1981, 41, 3615-3620.
114. Tennekes, H. A.; Elder, L.; Kunz, H. W. Carcinogenesis
     1982, 80, 397-408.
115. Krzystynik, K.; Hugo, P.; Flipo, D.; Fournier, M. F.
     Toxicol. Appl. Pharmacol. 1985, 80, 397-408.
116. Cabral, J. R. P.; Hugo, P.; Flipo, D.; Fournier, M. F.
     Nature (London) 1977, 191,363-366.
117. Shirai, T.; Miyata, Y.; Nakanishi, K.; Murasaki, G.; Ito.
     N. Cancer Lett. 1978, 4, 271-275.
118. Fuji, T.; Mikuriya, H.; Kamiya, N.; Hiraga, K. Food Cosm.
     Toxicol. 1986, 24, 207-211.
119. Abrahamsen, L. H.; Jerrkofsky, M. Appl. Env. Microbiol.
     1981, 41, 652-656.
120. Crocker, J. R. S.; Rozee, K. R.; Ozere, R. L.; Digout, S.
     C.; Hutzinger, O. Lancet 1974, ii, 22-24.
121. Nixon, J. E.; Koller, L. D.; Exon, J. H. J. Natl. Cancer
     Inst. 1979, 63, 1057-1063.
122. Fuji, K.; Epstein, S. S. Oncology 1979, 36, 105-112.
123. Borzsonyi, M.; Pinter, A. Neoplasia 1977, 24, 119-122.
124. Degrave, N.; Chollet, M. C.; Moutcshen, J . Environ.
     Health Persp. 1985, 60, 395-398.
125. Segall, Y.; Kimmel, E. C.; Dohn, D. R.; Casida, J. E. Mut.
     Res. 1985, 158, 61-68.
126. Decloitre, F.; Hamon, G. Mut. Res. 1980, 79, 185-192.
127. Garrett, N. E.; Stack, H. F.; Waters, M. D. Mut. Res.
     1986, 168, 301-325.
128. Cardy, R. H.; Renne, R. A.; Warner, J. W.; and Cypher, R.
     L. J. Natl. Cancer Inst. 1979, 62, 569-578.
129. Borzsonyi, M.; Pinter, A.; Surjan, A.; Torok, G. Cancer
     Lett. 1978, 5, 107-113.
130. Lijinksy, W. J. Env. Toxicol. Health 1984, 13, 609-614.
131. Cueto, C., Jr., Page, N. P., and Saffiotti, U. Roport on
     Carcinogenesis Bioassay of Technical Grade Chlordecone
     (Kepone ^{R}), Natl. Cancer Institute, DHEW Publication, NIH-
     76-1278 1978; 27pp.
132. Della-Porta, G.; Terracini, B.; Shubik, P. J. Natl. Cancer
     Inst., 1961, 26, 855-863.
133. National Cancer Institute. Report on carcinogenesis
     bioassy of chloroform. 1976.
134. Curtis, L. R.; Williams, W. L.; Mehendale, H. M. Toxicol.
     Appl. Pharmacol., 1979, 51, 283-293.

135. Hewitt, W. R.; Miyajima, H.; Cote, M. G.; Plaa;, G. L. Toxicol. Appl. Pharmacol., 1979, 48, 509-527.
136. Hewitt, L. A.; Caille, G.; Plaa, G. L., Can. J. Physiol. Pharmacol., 1986, 64, 477-482.
137. Purushotham, K. R.; Lockard, V. G.; Mehendale, H. M. Toxicol. Pathol., 1988, 16, 27-34.
138. Mehendale, H. M.; Purushotham, K. R.; Lockard, V. G., Exp. Mol. Pathol., 1989, 51, In Press.
139. Klingensmith, J. S.; Mehendale, H. M., Toxicol. Lett., 1982, 11, 149-154.
140. Agarwal, A. K.; Mehendale, H. M., Toxicology, 1988, 26, 231-242.
141. Agarwal, A. K.; Mehendale, H. M. Fundam. Appl. Toxicol, 1982, 2, 161-167.
142. Curtis, L. R.; Mehendale, H. M. Drug Metab. Dispos. 1980, 8, 23-37.
143. Davis, M. E.; Mehendale, H. M., Toxicology, 1980, 15, 91-103.
144. Klingensmith, J. S.; Mehendale, H. M. Drug Metab. Dispos. 1983, 11, 329-334.
145. Mehendale, H. M.; Klingensmith, J. S. Toxicol. Appl. Pharmacol. 1988, 93, 247-256.
146. Hewitt, W. R.; Miyajima, H.; Cote, M. G.; Plaa, G. L. Fed. Proc. 1980, 39, 3118-3123.
147. Charbonneau, M.; Ijima, M.; Cote, M. G; Plaa, G. L. Toxicology, 1985, 35, 95-112.
148. Glende, Jr., E. A.; Lee, P. Y. Exp. Mol. Pathol., 1985, 42, 167-174.
149. Dolak, J. A.; Britton, R. S.; Glende, Jr., E. A.; Recknagel, R. O. J. Biochem Toxicol. 1987; 2, 57-66.
150. Bell, A. N.; young, R. A.; Lockard, V. G.; Mehendale, H. M. Arch. Toxicol. 1988; 61, 392-405.
151. Kodavanti, P. R. S.; Joshi, U. M.; Young, R. A.; Bell, A. N.; Mehendale, H. M. Arch. Toxicol. 1989, 63, In Press.
152. Kodavanti, P. R. S.; Joshi, U. M.; Lockard, V. G.; Mehendale, H. M. J. Appl. Toxicol., 1989, 9, In Press.
153. Kodavanti, P. R. S.; Joshi, U. M.; Young, R. A.; Meydrech, E. F.; Mehendale, H. M. Toxicol. Pathol., 1989; 17, In Press.

RECEIVED July 13, 1989

Chapter 9

Biological Issues in Extrapolation

Raymond S. H. Yang[1], James Huff[1], Dori R. Germolec[1], Michael I. Luster[1], Jane Ellen Simmons[2], and John C. Seely[3]

[1]National Toxicology Program, National Institute of Environmental Health Sciences, P.O. Box 12233, Research Triangle Park, NC 27709
[2]Health Effects Research Laboratory, U.S. Environmental Protection Agency, Research Triangle Park, NC 27711
[3]Pathco Inc., P.O. Box 12796, Research Triangle Park, NC 27709

Approximately 41% (26/63) of the pesticides evaluated in the chronic toxicity and carcinogenicity studies of the National Toxicology Program (NTP) showed varying degrees of carcinogenicity. Since those chemicals nominated to the NTP for carcinogenicity studies usually represent a sampling of potentially "problem" chemicals, this ratio (i.e., 41%) does not implicate the actual percentage of carcinogenic chemicals among all pesticides. In general, results from epidemiological studies are of limited value in prevention and have been done on few pesticides. Furthermore, it is impossible to conduct chronic toxicity/carcinogenicity studies in humans. Therefore, laboratory animals must be utilized as surrogates, at least for the foreseeable future. To assess potential risks of chemicals to humans, extrapolation of the findings in laboratory animals to possible health effects in humans is inevitable. Several important biological issues must be considered in such a process; these include extrapolation between doses, species (including strain and sex), routes of administration and exposure regimens (e.g., intermittent vs constant rate). While there is no perfect surrogate for humans in the evaluation of carcinogenic potential and other toxicities of chemicals and other agents, rodents still represent the best models. Any imperfection in such a process (i.e., hazard identification and risk assessment of any given chemical or other agent) must be taken into consideration during extrapolation. The state-of-the-art technique or tools, such as physiologically based pharmacokinetics/computer modeling, should be considered and utilized judiciously to minimize the intrinsic imperfections in the process. The continuing refinement and improvement of the entire process (from selection of chemicals for testing, experimental design and conduct, interpretation and analysis of data, interpolation and extrapolation, to risk assessment) is essential as scientific advances are made in these areas. Other important issues related to hazard identification and risk assessment include multiple chemical exposure as well as multiple route exposure; these must be incorporated into the process as information and tools become available.

In a recent review article (1), the Council on Scientific Affairs of the American Medical Association reported their findings on the cancer risk of pesticides in agricultural workers. Their conclusions were: (a) The primary hazard of pesticide exposure is the development of acute toxic reactions as a result of dermal contact with or inhalation of a relatively large dose. The effects usually are manifested within minutes or hours of contact; (b) Epidemiological studies offer only limited evidence at best that pesticides may be carcinogenic; (c) A large number of pesticidal compounds have shown evidence of genotoxicity or carcinogenicity in animal and in vitro screening tests, but no pesticides -- except arsenic and vinyl chloride (once used as an aerosol propellant) -- have been proved definitely to be carcinogenic in man; (d) With few exceptions, the long term (e.g., carcinogenic) effects of pesticides on human health have been difficult to detect. Perhaps the health risks are sufficiently small that they are below the power of epidemiologic studies to detect. But it is also possible that there are very few effects to humans at all. While a comprehensive review of the epidemiologic studies conducted world-wide on pesticides, fertilizers, and/or agricultural practices (e.g., poultry, hog, dairy production) yielded conflicting results, the limitations of epidemiologic studies were also given (1) as shown in the following quotation.

"Epidemiologic analysis of data in man often fails to yield conclusive results because the size of the study group is too small (statistical power is inadequate to detect a difference), exposure data are lacking, concomitant or prior exposures to other known or suspected carcinogens may interact with the compound(s) in question and confound the analysis, or a proper control group cannot be identified. Then, too, some well-defined studies in different groups under different circumstances may yield totally conflicting findings."

Even if there were no such limitations, epidemiologic analysis is retrospective study; it cannot be depended upon for the detection and prevention of potential health hazards to the public. In comparison to epidemiologic studies, animal bioassays are short in duration, relatively inexpensive, easily perfomed under controlled conditions, and are reliable predictors for known human carcinogens. Therefore, from the point of view of identifying and preventing public health hazards, animal bioassay is an invaluable tool.

As of April 1989, the National Toxicology Program (NTP) and the National Cancer Institute (NCI; up to 1981), have completed chronic toxicity/carcinogenicity studies on 385 chemicals (2). Of the chemicals studied, 63 were considered pesticides (General or Unclassified) (Tables I and II). Approximately 41% (26/63) of the pesticides eva-

Table I. NCI/NTP Carcinogenicity Study Results for Pesticides: Carcinogenic Pesticides Under the Experimental Conditions

Chemical Name	TR No.	Route	Carcinogenicity Results			
			MR	FR	MM	FM
Aldrin	021	Feed	E	E	P	N
Captan	015	Feed	N	N	P	P
Chloramben	025	Feed	N	N	E	P
Chlordane	008	Feed	N	N	P	P
Chlorobenzilate	075	Feed	E	E	P	P
3-Chloro-2-methylpropene	300	Gav	CE	CE	CE	CE
Chlorothalonil	041	Feed	P	P	N	N
Daminozide	083	Feed	N	P	E	N
1,2-Dibromo-3-chloropropane (DBCP)	028	Gav	P	P	P	P
	206	Inh	P	P	P	P
1,2-Dibromoethane (EDB)	086	Gav	P	P	P	P
	210	Inh	P	P	P	P
1,4-Dichlorobenzene	319	Gav	CE	NE	CE	CE
1,2-Dichloropropane	263	Gav	NE	EE	SE	SE
1,3-Dichloropropene (Telone II)	269	Gav	CE	SE	IS	CE
Dichlorvos	342	Gav	SE	EE	SE	CE
Dicofol	090	Feed	N	N	P	N
Ethylene oxide	326	Inh			CE	CE
Heptachlor	009	Feed	N	E	P	P
Mirex	313	Feed	CE	CE		
Monuron	266	Feed	CE	NE	NE	NE
Nitrofen	184	Feed	N	N	P	P
	026	Feed	IS	P	P	P
Piperonyl sulfoxide	124	Feed	N	N	P	N
Sulfallate	115	Feed	P	P	P	P
Tetrachlorovinphos	033	Feed	N	P	P	P
Toxaphene	037	Feed	E	E	P	P
2,4,6-Trichlorophenol	155	Feed	P	N	P	P
Trifluralin	034	Feed	N	N	N	P

TR = NTP Technical Report; Gav = Gavage; Inh = Inhalation; MR = Male rats; FR = Female rats; MM = Male mice; FM = Female mice
For experiments evaluated by the NCI or the NTP prior to June 1983, results are reported as "positive" (P), "negative" (N), "equivocal" (E), or "inadequate" (IS). In June 1983, the NTP adopted the use of "categories of evidence": two of the five categories correspond to positive results ["clear evidence" (CE) and "some evidence" (SE) of carcinogenicity], one is for uncertain findings ["equivocal evidence" (EE)], one is for negative studies ["no evidence" (NE)], and one is for studies that cannot be evaluated because of major flaws ["inadequate studies" (IS)].

Table II.   NCI/NTP Carcinogenicity Study Results for Pesticides:
Non-carcinogenic Pesticides Under the Experimental
Conditions

Chemcial Name	TR No.	Route	Carcinogenicity Results			
			MR	FR	MM	FM
Aldicarb	136	Feed	N	N	N	N
Anilazine	104	Feed	N	N	N	N
Azinphosmethyl	069	Feed	E	N	N	N
Calcium cyanamide	163	Feed	N	N	N	N
2-Chloroethyltri-methylammonium chloride	158	Feed	N	N	N	N
Chloropicrin	065	Gav	IS	IS	N	N
Clonitralid	091	Feed	N	E	IS	N
Coumaphos	096	Feed	N	N	N	N
Diazinon	137	Feed	N	N	N	N
Dichlorvos	010	Feed	N	N	N	N
Dieldrin	021	Feed	N	N	E	N
	022	Feed	N	N		
1,2-Dichlorobenzene	255	Gav	N	N	N	N
Dichlorodiphenyl-dichloroethane (TDE)	131	Feed	E	N	N	N
Dichlorodiphenyl-trichloroethane (DDT)	131	Feed	N	N	N	N
Di(p-ethylphenyl)dichloroethane (DDD)	156	Feed	N	N	N	E
Dimethoate	004	Feed	N	N	N	N
Dioxathion	125	Feed	N	N	N	N
Endosulfan	062	Feed	IS	N	IS	N
Endrin	012	Feed	N	N	N	N
Fenthion	103	Feed	N	N	E	N
Fluometuron	195	Feed	N	N	E	N
Lindane	014	Feed	N	N	N	N
Malaoxon	135	Feed	N	N	N	N
Malathion	024	Feed	N	N	N	N
	192	Feed	N	N		
Methoxychlor	035	Feed	N	N	N	N
Methyl parathion	157	Feed	N	N	N	N
Mexacarbate	147	Feed	N	N	N	N
Parathion	070	Feed	E	E	N	N
Pentachloronitrobenzene	061	Feed	N	N	N	N
	325	Feed			NE	NE
o-Phenylphenol	301	SP			NE	NE
Phosphamidon	016	Feed	E	E	N	N
Photodieldrin	017	Feed	N	N	N	N
Picloram	023	Feed	N	E	N	N
Piperonyl butoxide	120	Feed	N	N	N	N
Rotenone	320	Feed	EE	NE	NE	NE
2,3,5,6-Tetrachloro-4-nitroanisole	114	Feed	N	N	N	N
Triphenyltin hydroxide	139	Feed	N	N	N	N

SP = Skin paint; For all other abbreviations, see Table I footnotes

luated showed varying degrees of carcinogenicity in animal studies
(2); these carcinogenic pesticides and those found to be non-
carcinogenic in the test systems are presented in Tables I and II,
respectively. The criteria used to determine whether a pesticide is
carcinogenic are according to the NTP or NCI convention; that is: one
or more of the four experiments (i.e., male rats, female rats, male
mice, female mice) showed "clear evidence", "some evidence" of carcinogenicity, or was ruled to be "positive" for carcinogenicity
(3,4). A caution must be made here that the percentage (i.e., 41%)
for pesticides tested positive as carcinogens does not mean that the
same percentage will hold true for all pesticides or all chemicals.
The fact that these chemcials were nominated to the NCI or NTP for
testing means there was concern that these chemicals could be potentially troublesome; thus the sampling is slanted toward potential
carcinogens.
    As shown in Figure 1, the primary efforts of the NTP center on
chemical nomination, evaluation of the existing information, study
design and conduct, data analysis and interpretation, and finally the
preparation, public peer-review, and publication of technical
reports. However, the information presented in the NTP Technical
Report is frequently used world-wide for hazard identification and
risk assessment, often leading toward regulatory decisions and
actions. Even though the NTP is not directly involved in the regulatory processes, the data generated at the NTP play an important role
in these processes.

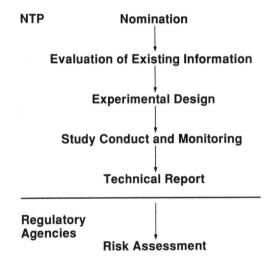

Figure 1. Flow chart of the functions of the NTP in the toxicologic characterization of chemicals and its role in relation to
regulatory agencies.

Once a chemical is demonstrated to have carcinogenic activity in laboratory rodents, extrapolations from "animal carcinogen" to "human carcinogen" and from "high dose" (i.e., the dose used in animal studies) to "low dose" (i.e., the dose humans are exposed to) are inevitable in the risk assessment process. Many biological issues are automatically brought into the limelight. At least some of these issues had been "locked in" as early as the experimental design stage. Those issues most often discussed are species, strain, sex, dose, route, and to some extent exposure scenario. Though important in the risk assessment process, the issues of toxicological interactions and the health effects of chemical mixtures have received comparatively little attention. In this paper, we will only discuss the current thoughts on a few selected issues with a special emphasis on toxicological interactions and chemical mixtures.

Species Extrapolation

Are rats and mice good models for humans toxicologically? This is an age-old question and opinions continue to be varied. But "Is a human a good model for humans?" Assuming there are no ethical and moral considerations and we, as a society, could perform toxicological studies including carcinogenicity bioassays in humans, would there be arguments? The answer is most likely a yes!, because there are so many environmental and life style differences (e.g., dietary preference, smoking, coffee drinking, alcohol indulgence, exercise, etc.) among individuals that the outcome of the results may easily be influenced. Assuming further that we built "human cages" with exercising machines, formulated "NIH-07 Human Chow," and completely eliminated all environmental and life style variables, would we then be more comfortable? The answer is no because there are known human pharmacogenetic differences such as fast acetylators and slow acetylators, and other biochemical and physiological diversities. There will always be arguments such as "Why didn't you use a different subpopulation?"

The point is that there is not and will never be a perfect system and that we have to settle for less than perfect. The right question to ask then is "Given the fact that there is no perfect system, how good are rats and mice as models for humans in toxicology?" Reviewing the literature regarding the development of the carcinogenicity bioassay, one finds such discussion as the one given by Shimkin (5) who proposed a general set of criteria for the selection of test animals in carcinogenicity studies: (a) availability; (b) economy; (c) sensitivity to carcinogens; (d) stable as to response; (e) similarity to human in regard to metabolism; (f) similarity to human in regard to pathology responses. When all these criteria are considered realistically, not too many laboratory animal species may be used for carcinogenicity studies. Other examples in the literature on the selection of species for carcinogenicity bioassay include Sontag (6) and Weisburger and Weisburger (7). On the basis of certain criteria, Sontag (6) suggested that the only

species suitable for large scale, long term carcinogenicity bioassays
are the rat, mouse, and hamster. Weisburger and Weisburger (7), on
the other hand, indicated that when specific questions are asked of
the bioassay rather than the simple endpoint of the increases of
neoplasms, a variety of species can be used. Thus, the NTP Ad Hoc
Panel on Chemical Carcinogenesis Testing and Evaluation suggested
examples where the specific relationship of a highly defined genetic
background and the induction of neoplasms in animals harboring few
oncogenic viruses may lead to the use of fish or insects (8). In our
opinion, considering the scientific as well as practical issues,
laboratory rodents are still the best models available to provide
information regarding the potential toxicological consequences of
chemicals in question. As responsible scientists in the public
health arena, we do not have the luxury to ignore the results from
animal studies even if they are less than perfect models for humans.
For instance, in recent years, there has been higher prevalence of
fin rot, tumors and other lesions in Winter Flounders caught in the
Boston Harbor (9). Are we to take such findings as a valid warning
from nature and consider them as the results of a "natural bioassay"
on the pollution in the area, or simply ignore it on the basis that a
fish is not a human?
     There is abundant debate in the literature about the use of
laboratory rodents as models for humans; a recent example appeared in
Statistical Science (10-16). Some scientists considered rodents very
good surrogates for humans in predicting carcinogenicity; for
instance, of the known human carcinogens, 86% to 100% (depending on
the data base used) were shown to be animal carcinogens (12,17-19).
Others (10), apparently using the same data base but only the
"sufficient evidence" category as defined by the International Agency
for Research on Cancer (IARC) (18) in their calculation, considered
rodents to be poor models for humans because the concordance was poor
(59%). At issue though is not whether or not the concordance is 100%
or 59% between human and animal models because even a selected sample
of humans probably do not have 100% concordance with the rest of the
human population. We know laboratory rodents are not perfect surrogates for humans but they definitely provide valuable scientific
information regarding the intrinsic biological activities of chemicals.
     It is important for scientists in toxicology to recognize the
imperfection of the animal bioassays and to try to continually develop better scientific methods to minimize the differences in the
extrapolation between species and between high and low doses. As a
case in point, the incorporation into the risk assessment process of
physiologically based pharmacokinetics/computer modeling which takes
into consideration physiological constants (e.g., body size, organ
and tissue volumes, blood flow, and ventilation rates), biochemical
constants (e.g., metabolic constants, partition coefficients for
blood, tissues and air), and mechanistic factors (e.g., target
tissues and metabolic pathways) of various species, including human,
represents a major scientific advance in recent years (20).

The importance of continuing refinement and improvement of the hazard identification and risk assessment process based on newly developed scientific information may be illustrated by another example using a pesticide, EDB or ethylene dibromide. Critics of cancer risk assessment based on animal bioassays frequently make reference to a paper by Ramsey et al. (21). In that paper (21), Ramsey and coworkers questioned the validity of the EPA's estimate of 100% lifetime incidence of cancer to be expected in humans exposed to a concentration of 0.4 ppm (3.1 mg/m^3) of EDB for 40 years. As pointed out by Ramsey et al. (21), the reliability of such a prediction is dependent upon the validity of the carcinogenesis model (one-hit model in this case) as well as its extrapolation from rat bioassay data to human population. To test these hypotheses, Ramsey et al. (21) compared the incidence of cancer predicted by the EPA one-hit carcinogenic model with that observed in a group of 156 workers employed in the production of EDB. While the one-hit model predicted a total of 85 tumors above the normal background incidence (2.2 tumors expected based on a comparison with the U. S. white male mortality rates) in this group of workers, the actual tumor incidence rate of 8 tumors was observed (21). Ramsey et al. concluded that use of the one-hit model appears to result in "highly exaggerated" risk estimates. Their results and conclusions have been cited as evidence that extrapolations from animal bioassays to human real-world exposures are implausible. A more recent attempt by Hertz-Picciotto et al. (22), however, has provided counter arguments. When cancer risks for EDB among the cohort of workers used by Ramsey et al. (21) were estimated by fitting several linear nonthreshold additive models to data from gavage and inhalation animal bioassays (22), the predicted upper bound risks were within a factor of 3 of the observed cancer deaths (Table III). Thus, Hertz-Picciotto et al. (22) concluded that the previous overestimate of risk to workers occupationally exposed to EDB was due to failure to consider their age at start of exposure when extrapolating from an animal bioassay with an exceedingly short latency period. In the quotation from Hertz-Picciotto et al. (22) below, the spirit of continuing improvement and refinement of the hazard identification and risk assessment process is elegantly reflected:

"The field of carcinogenic risk assessment is in its infancy. The primitiveness of methodology echoes the lack of a clear theory of carcinogenesis. However, the gaps in knowledge and the uncertainties in methods do not constitute sufficient justification for abandoning efforts to provide the public with plausible upper bounds for cancer risks due to environmental chemical exposures. For a large number of such exposures, these estimates will necessarily be based on animal data. When quantified human exposure data are available and are related to cancer risk, these data can be useful either as a basis for extrapolation or as a standard for assessing the plausibility of risk estimates based on animal data alone."

Table III. Total Cancer Deaths Predicted by Several Models for EDB-Exposed Workers

	Observed cancer deaths	One-hit	Models fitted to gavage data				Model fitted to Inhalation data One-hit: terminal sacrifice tumors	
			Proportional hazards		Multistage with time-to-tumor	Multistage with variable dosing		
			Stomach	Lung			Omitted	Included
0.9 ppm Exposure								
Texas	3	38.6	10.4	31.4	7.7	7.1	3.8	4.0
Michigan	5	21.2	5.9	13.0	3.5	4.5	2.3	2.4
Overall	8	59.8	16.3	44.4	11.2	11.6	6.1	6.4
3 ppm Exposure								
Texas	3	56.6	21.4	50.4	15.5	14.0	4.2	4.9
Michigan	5	34.2	11.7	24.3	6.1	7.7	2.5	2.9
Overall	8	90.8	34.5	74.7	21.6	21.7	6.7	7.8

Condensed from Hertz-Picciotto (22)

Maximum Tolerated Dose (MTD)

Since quantitative risk extrapolation is presented in another chapter, we discuss briefly here the issue of maximum tolerated dose (MTD). Actually "maximum tolerated dose" is a rather nebulous term without specific qualifiers. A generally accepted definition does not exist. The definition for MTD, given in one source (8), is "...this dose is determined by prechronic studies which aid in the identification of a dose level which, when given for the duration of the chronic study as the highest dose, will not impair the normal longevity of the treated animals from effects other than the induction of neoplasms. Such dose should not cause morphologic evidence of toxicity in organs other than mild changes such as slight hypertrophy or hyperplasia, inflammation or slight changes in serum enzymes..." In a second source (23), MTD is attributed to be the result of "...the simplistic approach of the National Cancer Institute's Bioassay Program, which is to conduct a 3 month range finding study with enough doses to find a level which suppresses body weight gain slightly, i.e., 10%. This dose is defined as the maximum tolerated dose (MTD) and is selected as the highest dose." In a third source (24), MTD is defined as "...a predictive dose obtained from analysis of subchronic study doses. In the oncogenicity study, the MTD should elicit toxicity without substantially altering the normal life span of the test species from effects other than tumor formation." In this case, it is of course really an exercise after the fact (i.e., after the completion of the chronic toxicity study) to evaluate if the doses selected were reasonable. Examples in such an evaluation for determining a MTD include body and organ weight effects as well as clinical and anatomic pathology (24). Thus, MTD must be considered in the context of the toxicological endpoints measured.

In the estimation of MTD and setting doses for chronic toxicity/carcinogenicity studies, one often hears advice against selecting any dose within the range of "saturation kinetics" (or "nonlinear kinetics"). The underlying reason for such advice is that we, the experimenters, should not "overwhelm" the animals' capacity of handling chemical insults. This is once again debate or argument resulting from nebulous terms. If one considers that any toxicity or toxicological response(s) is a manifestation of the disruption of homeostasis, then somewhere in the body there must be a system(s) (e.g., sites of action, transport mechanisms, metabolic pathways, binding sites, repair mechanisms, etc.) being overwhelmed. In fact, it has been suggested that toxicology is nonlinear pharmacokinetics (25). Therefore, at the stage of selecting doses for chronic toxicity studies based on prechronic toxicity data, it is very difficult to talk about nonlinear pharmacokinetics or saturation kinetics without defining precisely what parameters are involved.

As to the debate on the use of estimated MTD in chronic toxicity/carcinogenicity studies, recent information may be found in two letters by Abelson (26) and McConnell (27). Abelson suggests that the public has been misinformed through "a media barrage" by the results of chemical carcinogenesis studies in animals, particularly in rodents (26). One of the criticisms from Abelson was the use of "massive doses" (i.e., MTD) which, in his opinion, "vastly exceed those to which humans are likely to be exposed" (26). McConnell

(27), on the other hand, pointed out that, in at least three important instances (asbestos, benzene, and 1,3-butadiene), past occupational exposures are not different from the levels used in laboratory
experiments. Further, McConnell (27) defended the use of MTD in carcinogenicity studies because animals are able to detoxify chemicals
much faster than humans do, and to get an equivalent dose to the
target tissue in such a case would require much more of the chemical
in animals. He illustrated his point by using cigarette smoke as an
example; it takes an unusually high exposure of cigarette smoke to
cause cancer in laboratory aniumals as compared to humans (27).
McConnell asked that if cigarette smoke were an "unknown," would one
assume tobacco was safe because the dose to animals was higher than
the average human receives? (27).

Toxicologic Interactions

Toxicologic interaction has been defined as "a circumstance in which
exposure to two or more chemicals results in a qualitatively or quan-
titatively altered biological response relative to that predicted
from the actions of the single chemicals independently. The
multiple-chemical exposures may be simultaneous or sequential in
time, and the altered responses may be greater or smaller in magnitude" (28). While most if not all of the known examples of toxicologic
interactions concerning pesticides (29) involve respective chemicals
at concentrations far above levels of environmental contamination,
recent findings have uncovered toxicologic interactions at low enough
concentrations to be relevant to environmental pollutions. Table IV
illustrates an example of toxicologic interaction other than the
classical instances of insecticide synergism (e.g., pyrethrin and
piperonyl butoxide). Klingensmith and Mehendale (30) and Mehendale
(31) reported probably the first case where a pesticide, Kepone, at
an environmentally realistic level (i.e., 10 ppm) caused a dramatic
increase (67-fold) in the acute toxicity of carbon tetrachloride.
Although this synergistic effect of Kepone is rather specific in that
close structural analogs such as mirex and photomirex do not share
this property (31), one wonders whether other chemical(s) or chemical
mixtures, at environmentally low levels, might cause similar
synergistic toxicity. Indeed, the NTP has obtained some preliminary
experimental results (presented later) which may shed light upon this
question.
        Toxicologic interactions may occur in chronic toxicity and car-
cinogenicity studies. For instance, Wong and colleagues (32)
demonstrated a profound enhancement of mortality, tumor incidences
and the shortening of latency period for neoplasms of a well-known
fumigant, EDB, by disulfiram (antabuse) in a chronic toxicity study.
As shown in Table V, the numbers and the types of tumors increased
significantly when EDB and disulfiram were given together. Of
course, the reason for such a synergistic effect was due to the
interference of metabolic degradation of EDB by disulfiram (32). The
relevance of this study, as was the original purpose for the investigation (32), is the concern of the potential synergistic hazard for
people occupationally exposed to EDB (e.g., production workers, pest
control applicators) who might be simultaneously participating in an
alcohol control program using antabuse as a therapeutic agent.

Table IV. Enhancement of Acute Toxicity of Carbon Tetrachloride by
Low Level Dietary Pretreatment of Kepone

Dietary Pretreatment	48-hr LD_{50} (ml CC14/kg)	Increase in Mortality
Control diet	2.8	--
Kepone (10 ppm) diet	0.042	67-fold

Condensed from Klingensmith and Mehendale (30) and Mehendale (31)

These examples above demonstrate yet another area of uncertainty
in the extrapolation of animal toxicity studies to the hazard identification and risk assessment of humans. Since the data base in
this area is extremely limited, particularly at chemical concentrations which are environmentally realistic, major endeavours in
this area are urgently needed.

Table V. Major Histopathological Findings in Rats Exposed to EDB or
EDB/Disulfiram (EDB+DS) in the EDB/Disulfiram
Interaction Study

	EDB		EDB+DS	
	Male	Female	Male	Female
No. of Animals Examined	46	48	48	45
Liver				
hepatocellular tumors	2	3	36*	32*
Mesentary or omentum				
hemangiosarcoma	0	0	11*	8*
Kidney				
adenoma and adeno-				
carcinoma	3	1	17*	7*
Thyroid				
follicular epithelial				
adenoma	3	1	18*	18*
Mammary				
all tumors	---	25	---	13*
Lung				
all tumors	3	0	9*	2
No. of Rats with Tumor	25	29	45*	45*
No. of Rats with				
Multiple Tumors	10	8	37*	32*

* $P < 0.05$

Condensed from Wong et al. (32). Disulfiram was given in diet at 0.05%
and EDB inhalation exposure was at 20 ppm. The control and DS alone
groups (not shown here) had, in general, very low or no tumor incidence with respect to these organs; the only exception was that the
DS alone group had statistically higher incidence in mammary tumors
than did the control group.

Toxicology of Chemical Mixtures

Human exposure to chemicals, be it occupational or environmental, is
rarely limited to a single chemical. Even in a strictly controlled
situation such as a production plant for a single chemical, the exposure of workers to a variety of other chemicals in food, drink, personal hygiene, etc. is part of daily life. As illustrated very
clearly in a recent government publication on human health and the
environment (33), "...Each of us is exposed daily to multiple chemical substances in our environments. The food we eat is a complex
mixture of chemical substances. Our drinking water contains hundreds
of chemicals, even when it is obtained from a municipal
"purification" facility. The air we breathe and the things we touch
likewise contain a variety of chemicals with an almost limitless
range of compositions. Moreover, our habits and lifestyle may add
other chemical exposures. Tobacco smoke, for example, contains
thousands of substances. Additionally, the American public spends
between $4 billion and $8 billion per year on self prescribed and
self-administered drugs for self-diagnosed illness..."
    In the last few years, the NTP has been interested in the toxi-
cology of a mixture of 25 frequently occurring groundwater contaminants derived from hazardous waste disposal sites. The diagram
in Figure 2 is a summary of the projects completed or ongoing at the
NTP and neighbouring institutes. The genesis of such a program, the
formulation of a 25-chemical (19 organics and 6 inorganics) mixture
in deionized water, and the analytical and initial animal toxicology
work have been reported (34-41). In most of our animal studies, the
highest dose level, as shown in Table VI, contains individual chemical concentrations close to those detected in the groundwater samples
near hazardous waste disposal sites. When this mixture is given to
female B6C3F$_1$ mice for 14 or 90 days, suppression of immune function
was seen in 3 of the parameters examined (38); the results are summarized in Table VII. First, the suppression of bone marrow stem
cell proliferation, as expressed by the number of colonies formed of
the granulocyte-macrophage progenitor cells is presented. Note the
lack of response in the paired-water control group in the 14-day study; this finding suggests that the immunotoxic responses seen had
little, if anything, to do with the reduction of water consumption in
the treated groups. A clear dose-response relationship was
demonstrated in the 90-day study results on the suppression of bone
marrow stem cell proliferation. The second immunological endpoint
affected is the suppression of antigen (sheep red blood cell)-induced
antibody forming cells (Table VII); similar results as discussed
above for the stem cell suppression are also evident for this endpoint. Three host resistance assays following challenge with infectious agents (Listeria monocytogens, PYB6 syngeneic tumor cells, or
Plasmodium yoelii) were investigated (38). Altered resistance, as
expressed by % parasitemia, occurred in the group challenged with
Plasmodium (Table VII). These results collectively suggest that
long-term exposure to heavily contaminated groundwater may represent
a risk to the immune system (38).
    In another experiment (Figure 3), a collaborative effort between
the EPA and the NTP, we examined the effects of pretreatment with the
25-chemical mixture of groundwater contaminants for 14 days on the
hepatotoxicity of carbon tetrachloride in male Fischer 344 rats (39).

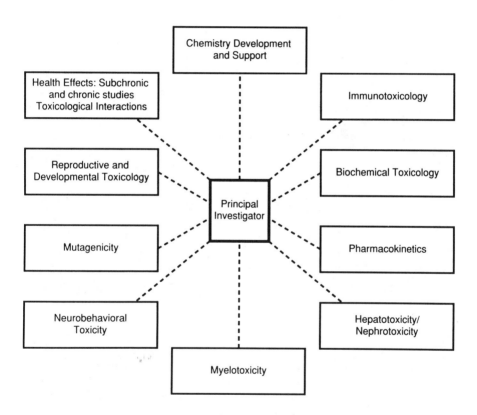

Figure 2.   A program of toxicology of complex chemical mixtures of groundwater contaminants at the NTP.

Table VI. Comparison of Concentrations of the Components of the
25 Chemical Mixture Dosing Solution with EPA
Survey Results

	A	B	
	Ave EPA Survey ppm	High Dose Concentrations (ppm)	Ratio B/A
Acetone	6.9	53	7.7
Arochlor 1260	0.21	0.01	0.05
Arsenic	30.6	9	0.29
Benzene	5.0	12.5	2.5
Cadmium	0.85	51	60
Carbon tetrachloride	0.54	0.4	0.74
Chlorobenzene	0.1	0.1	1
Chloroform	1.46	7	4.79
Chromium	0.69	36	52.2
DEHP	0.13	0.015	0.12
1,1-Dichloroethane	0.31	1.4	4.52
1,2-Dichloroethane	6.33	40	6.32
1,1-Dichloroethylene	0.24	0.5	2.08
1,2-trans-Dichloroethylene	0.73	2.5	3.42
Ethylbenzene	0.65	0.3	0.46
Lead	37.0	70	1.89
Mercury	0.34	0.5	1.47
Methylene chloride	11.2	37.5	3.35
Nickel	0.5	6.8	13.6
Phenol	34.0	29	0.85
Tetrachloroethylene	9.68	3.4	0.35
Toluene	5.18	7	1.35
1,1,1-Trichloroethane	1.25	2	1.6
Trichloroethylene	3.82	6.5	1.7
Xylenes	4.07	1.6	0.39

3 chemicals	> 10X	Average EPA Survey Concentration
14	1X - 10X	
8	< 1X	

As Table VIII demonstrates, the deionized water control group and the
drinking water chemical mixture alone group showed no histopathological changes in the liver. While the CCl_4 alone group showed a mild
centrilobular vacuolar degeneration at the dose level applied (0.075
ml/kg), the mixture/CCl_4 group showed, in addition to the mild
centrilobular vacuolar degeneration, minimal centrilobular hepatocellular necrosis (Table VIII). The necrotic changes in the liver in
the mixture/CCl_4 group also coincided with the elevation of serum
aspartate aminotransferase (177% of controls) and alanine aminotransferase (273% of controls). We are currently conducting an experiment
to determine the effect of prior exposure to the 25-chemical mixture
on the CCl_4 dose-response curve and to determine the influence of the
observed decrease in water and feed consumption on the apparent
enhancement of CCl_4 hepatotoxicity.

Table VII.   Immune Functions in B6C3F$_1$ Mice Affected by Exposure to a Chemical Mixture of Groundwater contaminants

Exposure Level (% Stock)	CFU-GM /10^5 cells[1]	PFC/Spleen (X 10^3)[2]	P. yoelii % parasitemia[3]
		14-day study	
0	59.8 + 6.0	172 + 16	14.2 + 1.2
0.2	57.2 + 1.3	231 + 16	9.2 + 1.0
2.0	55.4 + 3.2	157 + 24	19.8 + 2.9
20.0	37.8 + 0.4**	96 + 16**	22.1 + 2.9*
Paired water	50.5 + 1.3	172 + 17	N.D.
		3-month study	
0	55.3 + 2.1	189 + 49	10.1 + 1.5
1.0	52.6 + 2.9	120 + 19	9.4 + 2.9
5.0	43.5 + 4.6*	144 + 21	12.7 + 1.9
10.0	29.6 + 1.7**	93 + 15	20.8 + 2.9*

[1] The CFU-GM, granulocyte-macrophage colonies, were assayed by incubating femoral bone marrow cells in the presence of mouse lung conditioned medium as a colony stimulating factor at 37°C in 5% CO_2 for 7 days. Colonies of >50 cells were enumerated using a stereomicroscope. Values given represent mean + SE of CFU-GM per 10^5 cells for at least five mice per group.

[2] The antibody response to sheep erythrocytes was determined by enumerating plaque-forming cells (PFC) in splenic lymphocytes 4 days after primary immunization. Values given represent mean + SE of PFC per spleen for at least five mice per group.

[3] Infection with the malarial parasite P. yeolli was determined by quantitating the percent parasitemia on days 10, 12, and 14 following injection of 10^6 parasitized erythrocytes. Only peak day, day 12, of infection is shown. Values given represent mean + SE of eight mice per group.

N.D. = not done;
** Significantly different from control at P < 0.01;
* Significantly different from control at P < 0.05

SOURCE: Data are from ref. 38.

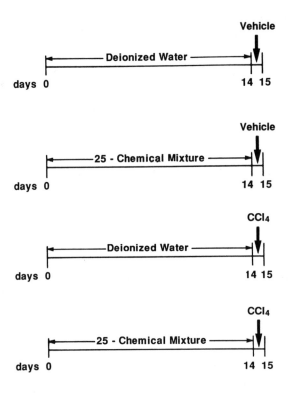

Figure 3. Experimental design of the toxicologic interaction study between a chemical mixture of 25 groundwater contaminants and carbon tetrachloride.

Table VIII. Enhancement of Carbon Tetrachloride Hepatotoxicity in Male Fischer 344 Rats by Prior Exposure to a Mixture of 25 Groundwater Contaminants

Treatment[1] Groups	No. Animal Examined	Clinical Chemistry Parameters		Liver Histopathology (No. Rats with Lesion)		
		AST[2]	ALT[2]	Normal	Vacuolar Degeneration	Cellular Necrosis
Control/ Control	4	52.8 ± 4.3	35.0 ± 4.1	4	0	0
Low Mixture/ Control	4	46.5 ± 2.6	32.8 ± 1.7	4	0	0
High Mixture/ Control	4	53.8 ± 6.2	39.0 ± 2.2	4	0	0
Control/ CCl4	4	55.0 ± 3.9	47.0 ± 7.5	0	4	0
Low Mixture/ CCl4	4	52.5 ± 4.0	41.2 ± 7.4	0	4	0
High Mixture/ CCl4	4	93.5 ± 36.5**	95.5 ± 49.7**	0	4	3

1 The rats were given deionized water (Control) or 1% (Low Mixture) or 10% (High Mixture) mixture stock for 14 days and then dosed by gavage a single dose of corn oil (Control) or CCl_4 at the rate of 0.075 ml/kg in corn oil; the animals were then sacrificed 24 hours later.

2 AST = Aspartate aminotransferase; ALT = Alanine aminotransferase

**Significantly different from the Control/Control group, the High Mixture/Control group, and the Control/CCl_4 group, $p < 0.01$.

The primary hazard of pesticide exposure is acute toxicity resulting from dermal contact with or inhalation of a relatively large dose (1). However, the toxicology of chemical mixtures at the level of environmental contamination will probably not involve acute toxic responses. It is most likely an insidious effect(s) disrupting the homeostasis of the organism. The exposed animals may appear totally "normal" clinically or based on conventional toxicological endpoints. However, such a subclinical state may provide a basis for enhancement or potentiation of otherwise mild toxic responses from an acute exposure(s) of chemical, physical, and/or biological agents. In this sense, the concept of a generic "promotor" or "enhancer" for any possible toxicity may be advanced for the potential toxicologic consequence of a mixture of environmental pollutants. These findings also raise the possibility for synergistic interaction between a background long-term, low-level chemical mixture exposure and an subsequent acute dose resulting from accidental exposure or drug intake including alcohol abuse. All these issues are not only relevant but very important in the extrapolation between animals and humans.

Concluding Remarks

Two years ago, in the ACS symposium on "Pesticides: Minimizing the Risks", the utilization of physiologically based pharmacokinetics/computer modeling for dealing with the complex issues of extrapolation, toxicological interaction, and chemical mixtures was suggested (42,43). As shown in Figure 4, the Stage I effort has already been

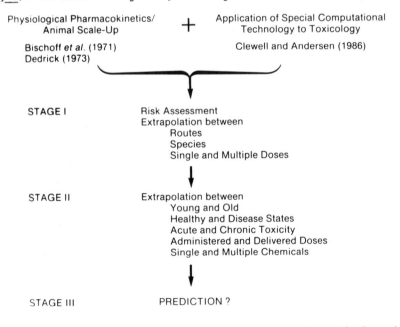

Figure 4. A suggested approach utilizing physiologically based pharmacokinetics and computer technology for the extrapolation and prediction of various situations in toxicology. (Reproduced from ref. 42. Copyright 1987 American Chemical Society.)

in progress. Some scientific activities are also evident for the effort of Stage II and beyond (44-49). In the last two years, a great deal of progress has been made in this area including a Workshop on "Pharmacokinetics in Risk Assessment" organized by the National Research Council which resulted in the publication of volume 8 of the Drinking Water and Health series (20). However, like any other new technique, the computer modeling of physiologically based pharmacokinetics and its application to risk assessment is going through an evolutionary phase. Much debate and controversy have been appearing in the literature. As a case in point, EPA's adoption of physiologically based pharmacokinetics in their risk assessment process and the related criticisms and debate reflect the current state of flux (48,49). Nevertheless, all these activities and debates will inevitably help to improve and refine this approach. Eventually, the scientific community and the public at large will have a better risk assessment method.

Acknowledgements

Some of the work mentioned in this report was supported in part or whole by funds from the Comprehensive Environmental Response, Compensation and Liability Act trust fund by interagency agreement with the Agency for Toxic Substances and Disease Registry, U. S. Public Health Service. Many colleagues in NIEHS and EPA helped in the studies mentioned in this paper; their effort and devotion are gratefully acknowledged. We thank Drs. E. E. McConnell (retired July 1988) and M. P. Dieter of NIEHS and Dr. D. M. DeMarini of EPA for reviewing the manuscript and for their helpful suggestions.

Disclaimer

The research described in this article has been reviewed by the Health Effects Research Laboratory, U. S. Environmental Protection Agency and approved for publication. Approval does not signify that the contents necessarily reflect the views and policies of the Agency nor does mention of trade names or commercial products constitute endorsement or recommendation for use.

Literature Cited

1. Council on Scientific Affairs, American Medical Association. JAMA 1988, 260, 959-66.
2. Chemical Status Report, National Toxicology Program, April 5, 1989.
3. Huff, J. E.; McConnell, E. E.; Haseman, J. K.; Boorman, G. A.; Eustis, S. L.; Schwetz, B. A.; Rao, G. N.; Jameson, C. W.; Hart, L. G.; Rall, D. P. Ann. N. Y. Acad. Sci. 1988, 534, 1-30.
4. Haseman, J. K.; Huff, J. E.; Zeiger, E.; McConnell, E. E. Environ. Health Perspect. 1987, 74, 229-235.
5. Shimkin, M. B. In Carcinogenesis Testing of Chemicals; Golberg, L., Ed.; CRC Press Inc., Cleveland, OH, 1974, p. 15.
6. Sontag, J. M. In Origins of Human Cancer; Book C Human Risk Assessment, Hiatt, H. H.; Watson, J. D.; Winsten, J. A., Eds.; Cold Spring Harbor Laboratory, 1977, pp.1327-1338.

7. Weisburger, J. H.; Weisburger, E. K. Methods in Cancer Res. 1967, 1, 307-399.
8. Report of the NTP Ad Hoc Panel on Chemical Carcinogenesis Testing and Evaluation, National Toxicology Program, August 17, 1984.
9. Murchelano, R. A.; Wolke, R. E. Science 1985, 228, 587-9.
10. Freedman, D. A.; Ziesel, H. Statistical Sci. 1988, 1, 3-28.
11. Breslow, N. Statistical Sci. 1988, 1, 28-33.
12. Haseman, J. K. Statistical Sci. 1988, 1, 33-9.
13. Moolgavkar, S. H.; Dewanji, A. Statistical Sci. 1988, 1, 39-41.
14. Kaldor, J.; Tomatis, L. Statistical Sci. 1988, 1, 41-3.
15. DuMouchel, W. Statistical Sci. 1988, 1, 43-4.
16. Freedman, D. A.; Ziesel, H. Statistical Sci. 1988, 1, 45-56.
17. Tomatis, L. Ann. Rev. Pharmacol. Toxicol. 1979, 19, 511-30.
18. Chemicals, Industrial Processes and Industries Associated with Cancer in Humans. Monograph Supplement 4. International Agency for Research on Cancer, 1982.
19. Overall Evaluation of Carcinogenicity: An Updating of IARC Monographs Volumes 1 to 42. Supplement 7. International Agency for Research on Cancer, 1987.
20. Pharmacokinetics in Risk Assessment, Drinking Water and Health Vol. 8, National Research Council, National Academy of Sciences, 1987.
21. Ramsey, J. C.; Park, C. N.; Ott, M. G.; Gehring, P. J. Toxicol. Appl. Pharmacol. 1978, 47, 411-4.
22. Hertz-Picciotto, I.; Gravitz, N.; Neutra, R. Risk Analysis 1988, 8, 205-14.
23. Stevens, K. R.; Gallo, M. A. In Principles and Methods of Toxicology; Student Edition, Hayes, A. W., Ed.; Raven Press, New York, 1984, p. 54.
24. Position Paper on Maximum Tolerated Dose (MTD) in Oncogenicity Studies, Environmental Protection Agency, April 1986.
25. Gehring, P. J.; Young, J. D. In Proceedings of the First International Congress on Toxicology; Toxicology as a Predictive Science, Plaa, G. L.; Duncan, W. A. M.; Eds.; Academic Press, New York, 1978, pp. 119-141.
26. Abelson, P. H. Science 1987, 237, 473.
27. McConnell, E. E. Science 1987, 238, 259-60.
28. Principles of Toxicological Interactions Associated with Multiple Chemical Exposures, National Research Council, National Academy Press, 1980.
29. Murphy, S. D. In Casarett and Doull's Toxicology; The Basic Science of Poisons, 3rd Edition, Klaassen, C. D.; Amdur, M. O.; Doull, J.; Eds.; Macmillan Publishing Co., New York, 1986, pp. 519-581.
30. Klingensmith, J. S.; Mehendale, H. M. Toxicol. Lett. 1982, 11, 149-154.
31. Mehendale, H. M. Fundam. Appl. Toxicol. 1984, 4, 295-308.
32. Wong, L. C. K.; Winsteon, J. M.; Hong, C. B.; Plotnick, H. Toxicol. Appl. Pharmacol. 1982, 63, 155-165.
33. Human Health and the Environment Some Research Needs, Report of the Third Task Force for Research Planning in Environmental Health Science, U. S. Department of Health and Human Services, NIH Publication No. 86-1277, 1984, pp. 147-164.
34. Yang, R. S. H.; Rauckman, E. J. Toxicology 1987, 47, 15-34.

35. Yang, R. S. H.; Goehl, T. J.; Brown, R. D.; Chatham, A. T.; Arneson, D. W.; Buchanan, R. C.; Harris, R. K. Fundam. Appl. Toxicol. 1989, In press.
36. Yang, R. S. H.; Goehl, T.; Jameson, C. W.; Germolec, D.; Luster, M. I.; Chapin, R.; Morrissey, R. E.; Schwetz, B. A.; Harris, R.; Chatham, A.; Arneson, D.; Moseman, R.; Collinsworth, N.; Bigelow, D. The Toxicologist, 1989, 9, 216.
37. Chapin, R. E.; Phelps, J. L.; Schwetz, B. A.; Yang, R. S. H. Fundam. Appl. Toxicol. 1989, In press.
38. Germolec, D. R.; Yang, R. S. H.; Ackermann, M. F.; Rosenthal, G. J.; Boorman, G. A.; Blair, P.; Luster, M. I. Fundam Appl. Toxicol. 1989, In press.
39. Simmons, J. E.; Yang, R. S. H.; Seely, J. C.; Svendsgaard, D.; McDonald, A. The Toxicologist 1989, 9, 58.
40. Schwetz, B. A.; Yang, R. S. H. In Proceedings on the Symposium on Experimental and Epidemiologic Applications to Risk Assessment of Complex Mixtures, Espoo, Finland, May 14-17, 1989, In press.
41. Shelby, M. D.; Tice, R. R.; DeMarini, D. M.; Yang, R. S. H. In Proceedings on the Symposium on Experimental and Epidemiologic Applications to Risk Assessment of Complex Mixtures, Espoo, Finland, May 14-17, 1989, In press.
42. Yang, R. S. H. In Pesticides. Minimizing the Risks; Ragsdale, N. N.; Kuhr, R. J., Eds.; ACS Symposium Series 336, American Chemical Society: Washington, DC, 1987, pp. 20-36.
43. Yang, R. S. H. CHEMTECH 1987, 17, 698-703.
44. Research to Improve Health Risk Assessments (RIHRA) Program, Environmental Protection Agency, June 1988.
45. Connally, R. B.; Reitz, R. H.; Clewell, III, H. J.; Andersen, M. E. Toxicol. Lett. 1988, 43, 189-200.
46. Connally, R. B.; Reitz, R. H.; Clewell, III, H. J.; Andersen, M. E. Comments Toxicol. 1988, 2, 305-19.
47. Selected Issues in Risk Assessment, Drinking Water and Health Vol. 9, National Research Council, National Academy of Sciences, 1988.
48. Perera, F. Science 1988, 239, 1227.
49. Moore, J. A. Science 1988, 240, 1125.

RECEIVED June 28, 1989

Chapter 10

Quantitative Risk Assessment

C. J. Portier

Division of Biometry and Risk Assessment, National Institute of Environmental Health Sciences, Research Triangle Park, NC 27709

Estimating risks from long-term chemical exposures at low dose levels represents a statistical and mathematical challenge with special relevance to environmental research. Choosing an adequate model for estimating the relationship between the administered dose and the tumor response is critical to reducing potential bias in the risk estimation process. This chapter will discuss the various assumptions and models used in carcinogenic risk assessment. The emphasis will be on the need to determine the shape of the dose–response relationship as well as the magnitude of the potential risk.

Estimation of Carcinogenic Risk

The standard procedure for estimating carcinogenic risks from exposure to chemicals is outlined in Table 1 and illustrated in Figure 1. The first step is to review the evidence concerning the mechanism of carcinogenicity of a particular chemical to get some idea concerning what forms to use for the two classes of mathematical models needed. The first class of models, commonly referred to as "tumor incidence models", relates the exposure dose of the chemical to the probability of cancer. Recently, this class of models has been characterized using hypothesized mechanisms through which normal cells become malignant. However, the second class of models, commonly referred to as "species conversion models", is beginning to receive a much greater degree of attention. These models concern species differences in the relationship between the administered dose of the chemical compound and the active toxicant which induces carcinogenesis. This active toxicant may be the parent compound, some metabolite of the parent compound or some other derivative of the compound. The risk assessor should have some knowledge of these mechanisms prior to attempting to quantify the carcinogenic risks. In practice, we frequently estimate carcinogenic risk without any such knowledge using conservative tumor incidence models and administered dose. When possible, the next step is to actually estimate the relationship between the administered dose (D) and the

**TABLE 1:   Basic Steps In Estimating A Safe
Exposure Level**

1. Consider Probable Mechanisms
  A. Utilize available knowledge to determine
      equivalent dose (ED) or some dose surrogate
  B. Determine appropriate tumor incidence model
2. Estimate the relationship between the administered
   dose and equivalent dose in the experimental
   animal
3. Express the administered doses in the
   carcinogenicity experiment in units of the
   equivalent dose
4. Estimate parameters of the tumor incidence model
   as a function of the equivalent dose
5. Determine a safe equivalent dose level
6. Estimate the relationship between the equivalent
   dose and the exposure dose in humans.
7. Express the safe equivalent dose in units of
   exposure dose in humans.

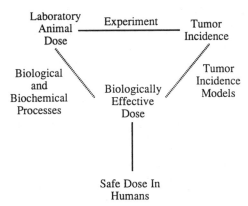

Figure 1:   Determination of Safe Exposure Levels from
Cancer Bioassay Data

"equivalent dose" (ED) in the test species (EPA, 1986). For the purposes of converting risks observed in one species to risks for another, it is assumed that the relationship between the ED and the probability of cancer is the same in all species. This assumption is critical to the estimation of carcinogenic risks since it provides the framework making species conversion a logical exercise (that is, if there were no equivalent toxic compound in both species, there is no logical reason to believe that tumor induction in the test species will result in tumor induction in humans). The estimation of the ED has generated a considerable amount of interest because the relationship between D and ED may be nonlinear (Hoel, Kaplan and Anderson, 1983). Failure to correctly model a nonlinear relationship could lead to an overestimate or an underestimate of the carcinogenic risk, a problem we will discuss later in this paper.

Several measures have been proposed for estimating the ED. The simplest is to assume that the ED is equal to or proportional to the administered dose, in which case no conversion is needed. The EPA has been using the average daily dose dose per unit surface area as the ED for most of their risk assessments (EPA, 1986). This value is typically calculated by multiplying the administered dose expressed in average daily dose per unit body weight by body weight to the 1/3 power, since the average surface area is proportional to body weight raised to the 2/3 power. Some chemicals must be transformed to reactive metabolites which bind to DNA in order to initiate the carcinogenic process. In these cases, the level of DNA adducts would seem to be a reasonable choice for the ED (Hoel, Kaplan and Anderson, 1983). Another potential ED would be the increased rate of mitosis or cell turnover induced by a chemical, usually as a result of chemical mitogenesis. Swenberg, Richardson, Boucheron and Dyroff (1985) propose using a quantity called the "initiation index", which is the product of the cell replication rate and the level of DNA adduct formation to allow for both types of biologically effective doses. For any of these ED's, the parameters in models which relate the administered dose to the ED are estimated from experiments in pharmacology, physiology and biochemistry.

Once the relationship between the administered dose and the ED has been estimated for the experimental species, the doses given to animals in a carcinogenesis bioassay can be converted from administered dose to ED. The tumor count data from the animal carcinogenicity experiment is then used to estimate the relationship between the ED and tumor incidence using standard tumor incidence models. The most commonly used model is the linearized multistage model (Anderson, et al., 1983), which is a modified version of the multistage model of Armitage and Doll (1954). There are numerous other models that have been employed, some of which are based upon statistical arguments and some of which attempt to mimic biological theory (see eg Krewski and Brown, 1980).

The tumor incidence model is then used to estimate a safe ED defined as that equivalent dose which yields a negligible increase in risk for the tumor over the background risk. For the purposes of risk estimation, it is assumed that this safe ED represents the safe ED in humans. It remains only to convert this safe ED in humans into a safe human exposure level. As before, a model relationship is assumed between the ED and exposure dose in humans (usually the same conceptual model as was used in the animal species) and parameter estimates are obtained. In addition to obtaining parameter estimates from the fields previously mentioned, a considerable number of parameters are estimated as simple allometric formulae of parameters which are easily obtained in animals, such as body weight. The complexity of these models, the number of parameters and the number of experiments used to estimate the parameters contribute to the overall uncertainty in the safe dose estimates.

Tumor Incidence Models and Low-Dose Extrapolation

In quantitative risk assessment, many different tumor incidence functions have been proposed, most of which do an adequate job of fitting the data in the experimental dose range. However, such models may differ considerably in the low-dose region, which is usually the critical region for estimating safe doses. This is illustrated by Figure 2. In Figure 2, two different tumor incidence models are plotted along with a set of typical bioassay data. If a model is chosen for which the slope of the dose-response curve is positive at dose zero, then small increases in dose will result in proportionate increases in the risk of getting a tumor. Models of this type are referred to as "low-dose linear" models and are illustrated by the "Linear Model" in Figure 2. Models for which the slope of the dose-response curve is zero or negative at low doses would result in virtually no change in carcinogenic risk for small amounts of dose. These models are typically referred to as "nonlinear" models and the "Non-Linear Model" shown in Figure 2 is of this type. It is clear from Figure 2 that it is difficult to assess which model is more appropriate for these data. Yet, if we modify the y-axis to look at added-risk over background and consider the low-dose region of these dose-response curves (detailed in Figure 2), there is a dramatic difference in the two curves. The "Linear Model" predicts a substantially smaller safe dose than does the "Non-Linear Model".

The economic and social implications of an incorrect model choice can be tremendous. Employing a "non-linear" model when in truth the response is "linear" can result in an unacceptably large risk to the exposed population. Employing a "linear" model when response is "non-linear" could result in unnecessarily restricting or banning the use of a potentially beneficial product. Thus, a major concern in quantitative risk assessment is to accurately describe the shape of the dose-response curve in the low-dose region.

In recent years, research in quantitative risk assessment has focused on the role of mechanism in estimating safe exposure levels. This work has concentrated on mechanistic models of carcinogenesis which are characterized as having biologically interpretable parameters. Because some mechanisms are low-dose linear and some are not, it is hoped that knowledge of the carcinogenic mechanism and the proper choice of a mechanistic model could result in an improved estimate of the shape of the dose-response curve. To facilitate a discussion of these models, a simple four-stage model of carcinogenesis will be used (Anderson, 1987). This model is illustrated in Figure 3.

In the model in Figure 3, there are five cell types; normal cells (N), intermediate cells (I), malignant cells (M) and two types of damaged cells (damaged normal cells (D_N) and damaged intermediate cells (D_I)). Assuming this model, for a cell to become malignant, it must pass from the normal state through each of the intermediate states. The biological theory behind the model can be described in simple terms as follows. Normal cells are allowed to divide and die or differentiate via a simple birth and death process where β_1 represents the rate of cell division (births) and δ_1 represents the rate of cell death. Normal cells transform into initial damaged cells via some type of genetic aberration (e.g. formation of DNA adducts, single strand breaks, gene amplification, chromosomal translocation) at a rate denoted by μ_1. The genetic aberrations in initial damaged cells are assumed to pertain to a single strand and can be repaired at the rate ρ_2 returning to the normal state. Damaged cells are also allowed to divide and die via a simple birth and death process with rates β_2 and δ_2 respectively. When cell division occurs in an initial damaged cell, the DNA damage is fixed in one of the daughter cells resulting in the creation of a single intermediate cell. The other daughter cell was derived from the strand of DNA without damage and is thus, a normal cell.

In intermediate cells, it is assumed that the DNA damage can no longer be repaired so that the resulting mutation is irreversible. Intermediate cells follow a process identical to the normal cells' process where they can divide, die or be damaged. The rates of these events are β_3, δ_3 and μ_3 respectively. When intermediate cells are damaged, they become damaged intermediate cells which can undergo repair (reversion to the intermediate cell state), death/differentiation or birth (resulting in one intermediate cell and one malignant cell) at rates ρ_4, δ_4 and β_4, respectively.

Given a model of the type shown in Figure 3, our interest is in calculating the probability of one or more

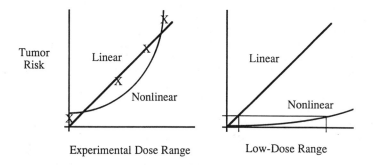

Figure 2:   Linear and Non-Linear Dose-Response Curves

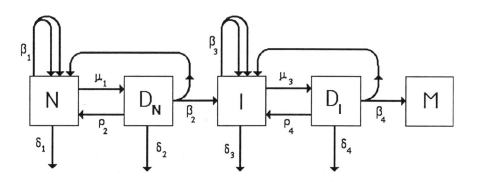

Figure 3:   A Four-Stage Model of Carcinogenesis With
Clonal Expansion and Repair

malignant cells by a specified time given a specified
exposure history. There exist mathematical tools (e.g.
Whittemore and Keller, 1978) and numerical tools (e.g.
Kopp and Portier, 1989) for determining this probability
from these types of models. It remains for us to consider
how exposure to a chemical carcinogen might alter this
time-course. Several mechanisms have been proposed for
chemical carcinogenesis which we will consider in the
context of this model, emphasizing the implications of the
assumed mechanism on the slope of the dose-response curve
at dose zero (i.e. low-dose linear mechanisms vs low-dose
non-linear mechanisms).
      One method by which chemicals may increase cancer
risks in animals is by modifying the rate of mitosis of
cells in a specific tissue or organ. This modification
could be due to numerous factors. One assumed mechanism
is the induction of regenerative hyperplasia in response
to tissue damage (e.g. Lewis and Adams, 1987). An
increase in the rate of mitosis of this type would most
likely represent an increase in the cellular birth rate,
β_i, for all stages of cell progression. It follows that
increasing the birth rate of initial damaged cells with no
subsequent increase in the repair rate will increase the
probability of the formation of an intermediate cell. The
same phenomenon would occur for final damaged cells ,
increasing the probability of seeing one or more malignant
cells. The relationship between dose and the increase in
the birth rates is generally thought to be non-linear,
possibly even having a threshold dose level below which
there is no increase in birth rate (see e.g. Swenberg et
al., 1987). However, the shape of the dose-response curve
in this situation is tied to two events; (i) the
relationship between dose of the chemical and the degree
of cytotoxicity (e.g. increases in the δ_i) and (ii) the
relationship between degree of cytotoxicity and the
increase in birth rates. These phenomenon need to be
studied in more detail before shape of the resulting dose
response curve can be determined.
      As mentioned earlier, the transformation from normal
cells to first-stage cells can result from damage to
cellular DNA which is not repaired prior to mitosis
(Barrett and Wiseman, 1987). A mutation could result from
numerous mechanisms such as single strand breaks to DNA or
the formation of DNA adducts.
      It is possible that some chemicals may specifically
alter the rate at which intermediate cells proliferate
without altering the proliferation rate of the other cell
types in the model. Thorslund, Brown and Charnley (1987)
contend that chemicals which act in this manner are
mitogens which increase the population of first-stage
cells through clonal expansion. Prehn (1964) suggests
other possible mechanisms resulting from selective
cytotoxicity. These mechanisms are also thought to relate
dose to response in a threshold-like manner and thus

treatment effects on these mechanisms might be modelled using a non-linear change in β_3 as a function of dose.

The final classification to be considered is chemicals which alter the transformation rate from initiated cells to malignant cells. Moolgavkar (1983) proposes one mechanism based upon the induction of homologous chromosome exchange during mitosis. Very little information exists on how chemicals might alter the rates for the second mutation, thus chemicals acting in this manner could result in either linear or nonlinear dose-response. The role of mutations in the carcinogenic process is discussed in detail in Barrett, 1987)

The model presented in Figure 3 allows for a variety of other effects such as the inhibition of DNA repair and/or multiple effects by a single chemical. It is unclear whether chemically induced changes in these rates would be linear or nonlinear in the low dose range.

There are numerous statistical problems with the use of mechanistic models in quantitative risk assessment. One question concerns whether a chemically induced increase in any of the rates in the model in Figure 3 are independent of the background rate or are adding to the existing rate. The answer to this question can have a serious effect upon the eventual shape of the dose-response curve (Hoel, 1980; Portier, 1987). Independent effects, even if they are linearly related to dose, can result in low-dose nonlinear behavior. On the other hand, additive effects are likely to be low-dose linear.

Another problem concerns the identifiability of carcinogenic mechanism using tumor incidence data. Using a simpler model than that shown in Figure 3, Portier (1987) examined the ability of tumor incidence data to accurately differentiate between a low-dose linear chemically induced increase in the rate of mutations from normal cells to initiated cells and a low-dose nonlinear chemically induced increase in the birth rate of initiated cells. It was shown that the probability of incorrectly classifying one mechanism as the other was quite high, exceeding 50% in some cases.

One final problem concerns the estimation of variability of carcinogenic risk estimates derived from a mechanistic model whose parameters are developed from multiple experiments. To consider all sources of variability in this context is a very difficult undertaking. An illustration of one possible approach is given by Portier and Kaplan (1989). In this case, a combination of bootstrapping techniques and Monte Carlo simulation techniques were used to resample from the observed data or to randomly sample from a distribution of possible values for a particular parameter. The example considered by Portier and Kaplan (1989) concerned a physiologically-based pharmacokinetic (PBPK) model for the distribution and metabolism of methylene chloride in mice and humans (Andersen, Clewell, Gargas, Smith and Reitz, 1987). This PBPK model was used by Andersen et al. to convert the administered dose into an equivalent dose

based upon the average area under the distribution-time
curve for a certain metabolite of methylene chloride.
The results of Portier and Kaplan indicate that there
could be a substantial increase in the variability of
safe-exposure estimates when using mechanistic models with
a large number of model parameters. Figure 4 illustrates
this result. The line labelled "Bioassay Data Only"
represents the frequency distribution of the safe human
exposure estimate when the only source of variability
considered is the variability of the tumor incidence data
from the animal carcinogenicity experiment on methylene
chloride. This is contrasted with the other line,
labelled "All Sources" where variability is accounted for
in all of the data used in the safe exposure estimates,
such as body weights, partition coefficients, metabolic
constants as well as the tumor incidence data from the
animal carcinogenicity experiment. It is clear there is a
substantial amount of variability unaccounted for by the
tumor incidence data since the range of the safe dose
estimates increases from less than one order to magnitude
to over 3 orders of magnitude. For more details on how
these distributions were derived, see Portier and Kaplan
(1989).

The increased variability noted by Portier and Kaplan
(1989) is a direct consequence of refining the modelling
process to allow all (or most) model parameters to vary
across individuals. This increased variability is not a
shortcoming of the use of these more complicated models.
Instead, it represents a more reasonable estimate of the
population variability with respect to the safe exposure
level, variability which should be estimated in order to

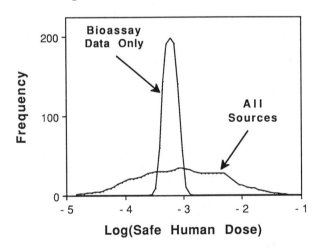

Figure 4: An Illustration of the Impact of Multiple
Sources of Variability on the Overall Distribution of the
Estimated Safe Exposure Level. (Data are from Portier and
Kaplan, 1989.)

know not only the mean safe exposure level, but the entire
distribution of safe exposures in the human population.

Summary

This paper has concentrated on the procedures used by
statisticians and mathematicians in estimating risks from
exposure to chemical carcinogens. A basic outline was
given in which the role of mathematics and statistics in
the risk estimation process was reviewed. One class of
mathematical models used in this context was studied in
detail; the multistage model of tumor incidence with
clonal expansion. There are numerous other mathematical
models used in the risk estimation process which include
other tumor incidence models, models relating administered
dose to the biologically effective dose, allometric
formulae, etc. In all cases, the current research effort
in statistics and mathematics is to use models which
presumably have a higher degree of biological plausibility
than those used previously. Because of this, these models
will also possess many of the problems discussed for the
multistage model in Figure 3; most notably: non-
identifiability of model parameters, inability to
distinguish linear tumor incidence data from non-linear
data in the low-dose region and difficulty in assessing
the overall variability of risk estimates derived from
multiple experiments.

References

Andersen, M., Clewell, H., Gargas, F., Smith, F., and
    Reitz, R. (1987). Physiologically based
    pharmacokinetics and the risk assessment process for
    methylene chloride. *Toxicology and Applied
    Pharmacology* 87: 185-202.
Anderson, E. and the Carcinogen Assessment Group (1983).
    Quantitative approaches in use to assess cancer risk.
    Risk Analysis 3: 277-295.
Anderson, M. (1987). Issues in biochemical applications
    to risk assessment: How do we evaluate individual
    components of multistage models? *Environmental
    Health Perspectives* 76, 175-180.
Armitage, P. and Doll, R. (1954). The age distribution of
    cancer and a multistage theory of cancer. *British
    Journal of Cancer* 8: 1-12.
Barrett, J. C. (1987) Genetic and epigenetic mechanisms
    in carcinogenesis. In Mechanisms of Environmental
    Carcinogenesis: Role of Genetic and Epigenetic
    Changes, Vol. I, pp. 1-15. J. C. Barrett, Ed. Boca
    Raton, CRC Press.

Barrett, C. and Wiseman, R. (1987). Cellular and
    molecular mechanisms of multistep carcinogenesis:
    Relevance to carcinogen risk assessment.
    Environmental Health Perspectives 76, 65-70.
EPA (1986). Guidelines for carcinogen risk assessment.
    51 Federal Register 33992, 1-17.
Hoel, D., Kaplan, N. and Anderson, M. (1983). Implication
    of nonlinear kinetics on risk estimation in
    carcinogenesis. Science 219: 1032-1037.
Hoel, D. (1980). Incorporation of background in dose-
    response models. Federation Proceedings 39: 73-75.
Kopp, A. and Portier, C. (1989). A note on approximating
    the cumulative distribution function of the time to
    tumor onset in multistage models. Biometrics, in
    press..
Krewski, D. and Brown, C. (1981). Carcinogenic risk
    assessment: A guide to the literature. Biometrics
    37, 353-366.
Lewis, J. and Adams, D. (1987). Inflammation, oxidative
    DNA damage and carcinogenesis. Environmental Health
    Perspectives 76, 19-28.
Moolgavkar, S. (1983). Model for human carcinogenesis:
    Action of environmental agents. Environmental Health
    Perspectives 50: 285-291.
Portier, C. (1987). Statistical properties of a two-
    stage model of carcinogenesis. Environmental Health
    Perspectives 76: 125-131.
Portier, C. and Kaplan, N. (1989). The variability of
    safe dose estimates when using complicated models of
    the carcinogenic process. A case study: Methylene
    chloride. Fundamental and Applied Toxicology, in
    press.
Prehn, R. (1964). A clonal selection theory of chemical
    carcinogenesis. Journal of the National Cancer
    Institute 32 (1): 1-17.
Swenberg, J., Richardson, F., Boucheron, J. and Dyroff, M.
    (1985). Relationships between DNA adduct formation
    and carcinogenesis. Environmental Health
    Perspectives 62: 177-183.
Thorslund, T., Brown, C. and Charnley, G. (1987).
    Biologically motivated cancer risk models. Risk
    Analysis 7: 109-119.
Whittemore, A. and Keller, J. (1978). Quantitative
    theories of carcinogenesis. SIAM Review 20, 1-30.

RECEIVED June 28, 1989

Chapter 11

Role of Structure–Activity Relationship Analysis in Evaluation of Pesticides for Potential Carcinogenicity

Yin-tak Woo and Joseph C. Arcos

Office of Toxic Substances, U.S. Environmental Protection Agency, Washington, DC 20460

Structure-activity relationship (SAR) analysis is essential for the development of pesticides and for the evaluation of cancer hazard and risk assessment. The critical factors that should be considered in SAR analysis and the profile of typical potent carcinogens are discussed. A scheme combining structural and functional criteria for suspecting chemical compounds of carcinogenic activity is presented. Selected classes of pesticides with carcinogenic potential are reviewed to exemplify structural and/or functional features responsible for their carcinogenic activity.

Structure-activity relationship (SAR) analysis is a critical tool in the research and development of new industrial and agricultural chemicals and is the first line of approach in the cancer hazard evaluation of chemicals. Careful SAR analyses can spot or reveal potential health hazard of new chemicals early in the research and development stage. SAR considerations are also essential for designing and selecting appropriate batteries of tests to study the potential toxicity of chemicals and to elucidate their molecular mechanisms of action.

There are various ways to approach SAR analysis. This chapter focuses on principles and concepts of mechanism-based SAR analysis along with an overview of the structural features and critical factors that should be considered in the evaluation of pesticides for potential carcinogenicity. Most of principles and concepts of SAR analysis covered in this chapter represent a distillate based on a series of monographs (1-5) and reviews (6,7) on the chemical induction of cancer by the authors. Readers are referred to these sources (particularly the Cumulative Index in ref. 5) for the original references of specific SAR studies.

Multi-stage and Multifactorial Nature of Chemical Carcinogenesis

The SAR analysis of carcinogens requires a basic understanding of their biochemical mechanisms of action. Chemical carcinogenesis may have many etiological factors and is a complex process that may be divided into at least three distinct stages: initiation (DNA damage which, if unrepaired or misrepaired, eventually leads to the formation of preneoplastic or "dormant" tumor cells), promotion (in which preneoplastic cells are expressed into individual tumor cells) and progression (involving progress to malignancy by histopathologic criteria). A variety of endogenous (host) factors such as immune competence, hormonal regulation, and exogenous factors such as diet, radiation, trauma/stress can play important contributory roles. Chemical carcinogens can exert their action by directly acting on these three stages of carcinogenesis as well as indirectly through the endogenous (host) factors.

Mechanistic Classification of Chemical Carcinogens

From the point of view of mechanism of action, carcinogens can be loosely classified as: (i) genotoxic carcinogens, and (ii) epigenetic carcinogens. Genotoxic carcinogens cause DNA damage directly, mostly through covalent binding to DNA, and are also called DNA-reactive carcinogens. Despite their structural variety, they have one feature in common -- they are either electrophiles per se or can be activated to electrophilic reactive intermediates. Epigenetic carcinogens do not bind covalently to DNA, do not cause DNA damage directly, and usually produce negative or inconsistent results in short-term tests for genotoxicity. They act by a variety of not clearly defined extrachromosomal mechanisms such as peroxisome proliferation, inhibition of intercellular communication, hormonal imbalance, cytotoxicity, etc. Prediction of their possible mechanism of action is important for meaningful SAR considerations.

Carcinogens which act as electrophilic reactants may be further classified as: (i) direct-acting carcinogens which are reactive as such, and (ii) indirect-acting carcinogens which require activation chemically (e.g., by acid or base), photochemically (e.g., uv) or metabolically. In general, direct-acting carcinogens tend to be locally active whereas those which require activation are usually carcinogenic mostly toward tissue(s) where activation occurs.

Critical Factors for SAR Consideration

There are four critical features that should be considered in SAR analysis: (i) physicochemical, (ii) molecular geometric, (iii) metabolic, and (iv) mechanistic.

   Physicochemical factors: Irrespective of its chemical structure or mechanism of action, the carcinogenic potential of a chemical compound is dependent on its physicochemical properties which determine its "bioavailability", that is, its ability to reach target tissues and cells. The most salient of these properties are:
   (1) Molecular weight: Compounds with very high molecular weight (over 1,000-1,500) and size have little chance of being absorbed in significant amounts; in general, they do not pose any substantial carcinogenic risk. There are of course important exceptions to the rule. High M.W. compounds which can be degraded in the gastrointestinal tract (e.g., by hydrolysis or microbial action) should be assessed considering their probable degradation products. Certain high M.W. polymeric substances (e.g., degraded carageenan) have local carcinogenic effect if ingested, inhaled or injected (see ref. 5).
   (2) Physical state: The physical state of a chemical compound may, to some extent, affect its capability to reach target tissues. Compounds which are highly volatile, or which can be inhaled as dust particles, may have "direct access" to nasopharyngeal and/or pulmonary tissues.
   (3) Solubility: In general, compounds which are highly hydrophilic are poorly absorbed and, if absorbed, are readily excreted. Thus, the introduction of hydrophilic groups (e.g., sulfonyl, carboxyl) into an otherwise carcinogenic compound usually mitigates and sometimes altogether abolishes its carcinogenic activity.
   (4) Chemical reactivity: Although many chemical compounds owe their carcinogenic activity to their electrophilic chemical reactivity, compounds which are "too reactive" are not carcinogenic. Compounds are considered "too reactive" if they hydrolyze or polymerize spontaneously and instantaneously, or react with noncritical cellular constituents before they can reach target tissues and react with key macromolecules. For example, the carcinogenic activity of the reactive

electrophile, ß-propiolactone can be abolished by the introduction of an exocyclic double bond which makes the compound (diketene) "too reactive" (see ref. 3). It is important to point out that the route of exposure is a key factor in considering whether the compound is "too reactive". For example, bis-chloromethylether may be considered "too reactive" if administered orally in aqueous solution ($t_{1/2}$ approximately 40 seconds), however, the compound is a potent nasal/pulmonary carcinogen if inhaled as vapor ($t_{1/2}$ in humid air may be as long as 25 hours) (see ref. 3).

Molecular geometric factor: Molecular size and geometry of a chemical compound affect its ability to reach target tissue and target macromolecules and its chance to be metabolically activated or detoxified. Many potent carcinogens/mutagens have a common feature -- they have a molecular size and shape favorable for intercalation into DNA plus a reactive or potentially reactive functional group. For example, it has been calculated (see ref. 1) that most potent carcinogenic polycyclic aromatic hydrocarbons have a planar structure with 4 - 6 rings (of which not more than four may be linearly connected) and a molecular size of between about 90 to 180 $\overset{\circ}{A}{}^2$. This size requirement coupled with the potential to be activated to electrophilic bay-region diolepoxide represent two critical factors that determine the carcinogenic activity of polycyclic aromatic hydrocarbons. The potent carcinogens found in overcooked food (e.g., Trp-P-1, P-2; Glu-P-1, P-2) are mostly planar tricylcic compounds (8) with amino group(s) that can be activated to electrophilic nitrenium ions. Many other potent carcinogens and mutagens (e.g., aflatoxin B_1, psoralen-8-glycidyl ether, acridine mustard, 2-acetylaminofluorene) are planar molecules with a favorable molecular size bearing an electrophilic functional group (see refs. 2-5).

Metabolic factor: Metabolism can both activate and detoxify chemical carcinogens. For direct-acting carcinogens, metabolism tends to decrease the activity. Many pesticides are direct-acting electrophiles. Thus, an estimate of the extent of metabolic detoxification of the pesticides is essential for accurate SAR prediction. For indirect-acting carcinogens, a delicate balance between the activation and detoxification pathways determines the carcinogenicity of the compound. Knowledge of the metabolic pathways of these chemicals can substantially enhance the accuracy of SAR analysis. For example, for carcinogenic polycyclic aromatic hydrocarbons, introduction of small substituents (such as methyl or fluoro) at the L-region increases the carcinogenicity, whereas any substitution at the site of activation (the bay-region benzo ring) decreases or annuls carcinogenicity (see refs. 1, 9, 15).

A variety of oxidative, reductive, hydrolytic and conjugating enzymes/enzyme systems activate chemical carcinogens (7). The mixed-function oxidases are by far the most well known activating system for most carcinogens (10). However, at least two other enzymes/enzyme systems deserve a special mention here. One is the relatively little known prostaglandin H synthase (PHS) that is increasingly considered to be the major activating system in many extrahepatic tissues (11). The PHS system has been shown to activate a variety of chemical carcinogens including some that were previously assumed to be "epigenetic" (see ref. 7). The herbicide, amitrole (3-amino-1,2,4-triazole), a thyroid carcinogen and goitrogen, for example, has been generally assumed to be an "epigenetic" carcinogen because of its goitrogenic activity and its inactivity in a variety of short-term mutagenicity assays (mostly using liver microsomes as the activating enzyme source). However, recent studies indicate that amitrole can induce gene mutation and in vitro cell transformation in Syrian hamster embryo fibroblasts which are known to contain PHS activity. Amitrole can be activated by microsomes from ram seminal vesicles (high in PHS activity) to reactive intermediates capable of binding covalently to DNA (12a). Another interesting enzyme is glutathione (GSH) S-transferase which is generally regarded to be a detoxifying enzyme. However, for dihaloalkanes such as the fumigant, 1,2-dibromoethane, GSH S-transferase is an activating rather than a detoxifying enzyme because the metabolite formed, a GSH conjugate, is actually a half sulfur mustard (GS-CH$_2$-CH$_2$-Cl) which can undergo cyclization to yield electrophilic episulfonium intermediate (see refs. 7, 12b).

Mechanistic factor: Depending on the projected mechanism of action, different approaches are needed in SAR analysis. However, since carcinogens often act by more than one mechanism, all possible mechanisms should be considered for evaluation of their applicability to the chemical in question.

Profile of Typical Potent Carcinogens

There are a number of characteristics that are common to most typical potent carcinogens:

(1) First, they must be able to reach target sites at sufficiently high level. As discussed earlier, the physicochemical properties of the chemical play a determining role. Reactive direct-acting electrophilic chemicals are often potent carcinogens if they are either

(i) volatile (e.g., bis-chloromethyl ether), (ii) administered by routes or conditions allowing direct access to target tissues, or (iii) resistant to metabolic detoxification (e.g., 1,2-dibromoethane which is activated rather than detoxified by GSH).

(2) Activation near or at the site of target is one of the most effective means of generating potent carcinogens. For this reason, the liver, the most important organ for metabolism, is a most frequently observed target organ for carcinogens that require metabolic activation. A number of carcinogens (e.g., 4-aminobiphenyl) can be activated to electrophilic intermediates in the liver, temporarily detoxified by forming conjugates (e.g., glucuronide), tranported in this "protected" form to a second target (e.g., urinary bladder), and reactivated by target-specific hydrolase (e.g., ß-glucuronidase) or acid hydrolysis (see refs. 2, 7, 12b).

(3) A reasonable lifetime to permit interaction with target macromolecules is also crucial. For reactive carcinogens, this means some ways to stabilize the reactive intermediate (see discussion under "SAR Consideration of Genotoxic Carcinogens"). For nonreactive carcinogens, this means metabolic stability so that the chemical may stay in the body longer.

(4) Selective, specific, and persistent interaction with DNA or other target macromolecules is another critical feature. For example, the potent carcinogen, aflatoxin B_1, after metabolic activation binds selectively to certain specific regions of DNA. The DNA adducts thus formed undergo post-binding conversion to adducts that are persistent and resistant to DNA repair enzymes (see ref. 5). Other potent carcinogens such as 7,12-dimethylbenz[a]anthracene and N-nitrosomethylurea have been shown to selectively bind to specific DNA regions that encode oncogenes (13). For 2,3,7,8-TCDD, it selectively binds to cytosolic Ah protein. Although the binding is not covalent, the persistent nature of 2,3,7,8-TCDD ensures mechanistically significant interaction (14).

(5) As discussed earlier, chemical carcinogenesis is a complex process involving multiple stages and many factors. Most potent chemical carcinogens exert effects on multiple steps or stages of carcinogenesis. The synergistic or complementary combination of these individual effects determines the potency of the carcinogen. For example, the potent carcinogen, benzo[a]pyrene, is metabolically activated to reactive bay-region diolepoxide that binds to DNA and initiates carcinogenesis (15), it is also metabolized to phenolic derivatives that act as promotors (16), and it is active as an immune suppressive agent (17).

SAR Consideration of Genotoxic Carcinogens

As discussed above, SAR analysis is more effective if we can predict/project the possible mechanism of action of the chemical. For genotoxic carcinogens, SAR analysis should include consideration of: (i) the nature of the electrophile present or to be formed, (ii) possibility of stabilization of the intermediate, and (iii) the molecule to which the electrophile is attached.

The electrophiles or electrophilic intermediates that are or are postulated to be responsible for the carcinogenic action of chemicals include: (i) positively charged carbonium, nitrenium, oxonium and episulfonium ions, (ii) free radicals, (iii) polarized double bonds, (iv) aldehydes, (v) strained rings such as epoxide, aziridine, lactones and sultones, and (vi) quinone/quinoid/quinoneimine structures. Based on their reactivity (Table I), electrophiles may be graded from "soft" to "hard" similar to the concept of "soft" and "hard" acids and bases (18). In general, soft electrophiles react preferentially with soft nucleophiles whereas hard electrophiles react preferentially with hard nucleophiles. Thus, since the nucleophilic sites in the purine and pyrimidine bases in DNA are moderately hard nucleophiles, moderately hard electrophiles tend to have the greatest likelihood of covalent binding to DNA. Soft electrophiles often deplete the cellular pool of noncritical soft nucleophiles (such as GSH) before they can react with DNA.

Even when activated near the target site, some electrophilic intermediates are still too reactive to travel from enzymic site to DNA. Resonance stabilization often provides the needed lengthening of lifetime for the reactive intermediate to reach its target macromolecules. A comprehensive study of the SAR of carcinogenic aromatic amines (2) best illustrates the point. Figure 1 shows the position(s) where attachment of an amine or amine-generating group (such as nitro, nitroso group) can yield carcinogenic aromatic amines. These positions correspond to the terminal end(s) of the longest conjugated system in the molecule. It is at these positions that the reactive nitrenium ions, generated by metabolic activation, can be most effectively stabilized by resonance with the aromatic ring system.

The molecular segment adjacent/attached to the electrophilic reactive site/group can affect carcinogenic potential in a variety of ways. In addition to molecular size and resonance stabilization, a number of other molecular parameters can modulate the carcinogenic potential of the electrophilic moiety. Attachment of normal cellular constituents or their structural analogs

TABLE I

Typical "Soft" and "Hard" Electrophiles and Their
Preferred Sites of Attack in Macromolecules[1]

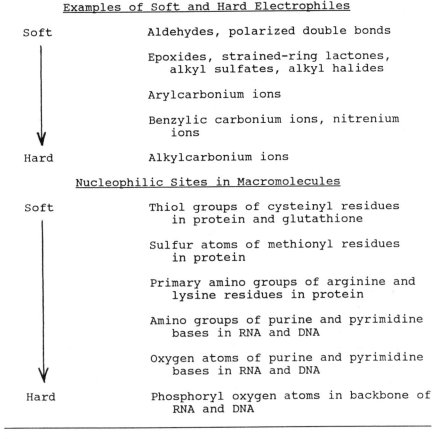

Examples of Soft and Hard Electrophiles

Soft   Aldehydes, polarized double bonds

                Epoxides, strained-ring lactones,
                    alkyl sulfates, alkyl halides

                Arylcarbonium ions

                Benzylic carbonium ions, nitrenium
                    ions

Hard   Alkylcarbonium ions

Nucleophilic Sites in Macromolecules

Soft   Thiol groups of cysteinyl residues
                    in protein and glutathione

                Sulfur atoms of methionyl residues
                    in protein

                Primary amino groups of arginine and
                    lysine residues in protein

                Amino groups of purine and pyrimidine
                    bases in RNA and DNA

                Oxygen atoms of purine and pyrimidine
                    bases in RNA and DNA

Hard   Phosphoryl oxygen atoms in backbone of
                    RNA and DNA

[1]Brian Coles, personal communication.

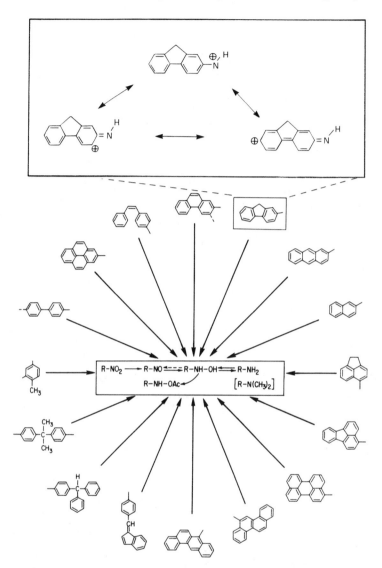

Fig. 1. Role of Resonance Stabilization in Contributing to Carcinogenic Activity of Aromatic Amines. The lower diagram indicates typical hydrocarbon moieties present in carcinogenic aromatic amines. The unconnected bond(s) in these moieties indicate the position where attachment of (an) amine or amine-generating group(s) yields carcinogenic compounds. These positions correspond to the terminal end(s) of longest conjugated system in the molecule and are the most favorable positions for resonance stabilization of reactive nitrenium ions generated by metabolic activation (<u>see</u> upper blocked diagram).

to the electrophilic moiety can often increase
carcinogenicity. In these instances the molecular segment
attached acts as a carrier to facilitate entry into the
target organ(s). For example, the pancreatic carcinogen,
streptozotocin, contains a N-nitroso moiety and a sugar
moiety. The former generates an electrophilic moiety
whereas the latter acts as a carrier (see ref. 5).
Attaching the reactive mustard group to uracil yields a
highly potent carcinogen (see ref. 3), presumably because
uracil can be readily taken up by cells. The location of
the electrophilic moiety(ies) in the molecule is also
important. Electrophilic moiety(ies) situated at the
terminal ends has(ve) a better chance of reacting with
target macromolecules than those situated at or
sterically "buried" in the middle of the molecule because
of steric hindrance. Molecules containing more than one
electrophilic moiety are more likely to be carcinogenic
than monofunctional ones particularly if the
electrophilic moieties are freely flexible and not too
far apart (see ref. 3).

SAR Consideration of Epigenetic Carcinogens

In contrast to genotoxic carcinogens, epigenetic
carcinogens act by a wide variety of mechanisms. Some of
the possible epigenetic mechanisms of chemical
carcinogenesis include: peroxisome proliferation,
inhibition of intercellular communication, microtubule
alteration, hormonal imbalance, cytotoxicity,
immunomodulation, inhibition of DNA methylation, etc.

    The evaluation of potential carcinogenicity based
on epigenetic mechanism, requires examination of the
following questions: (i) what is the most likely
mechanism(s)?, (ii) is information available to allow
SAR analysis based on this mechanism?, and (iii) are
possibly other mechanism(s) involved? For example, a
variety of peroxisome proliferators with different
chemical structures have been shown to be
hepatocarcinogenic (19). At least for the chlorinated
phenoxyacetic acids (20), some SAR can be discerned (Fig.
2). It appears that the degree of branching at the ω-1
carbon plus the electron-withdrawing capability of the
chlorinated phenoxy moiety contribute to the peroxisome
proliferative activity. Among the chlorinated
phenoxyacetic acid derivatives that have been tested for
carcinogenic activity, two of the more potent peroxisome
proliferators (ciprofibrate and clofibric acid) are
carcinogenic whereas for 2,4-D (a weak peroxisome
proliferator) there appears to be no convincing evidence
for carcinogenicity (21; see also refs. 3,4 and this
monograph). It remains to be studied whether this SAR of
peroxisome proliferative activity can be used to predict

Relative Peroxisome Proliferative Activity of Chlorinated Phenoxyacetic Acids and Related Compounds

Compound		R.P.*
Ciprofibrate	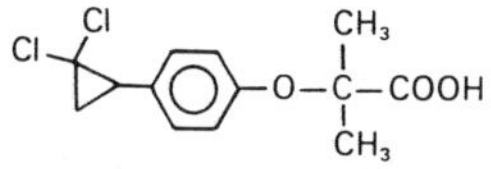 Cl-cyclopropyl-C_6H_4-O-$C(CH_3)_2$-COOH	20.3
Methyl clofenapate	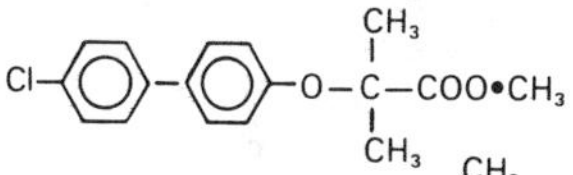 Cl-C_6H_4-C_6H_4-O-$C(CH_3)_2$-COO•CH_3	12.4
Clobuzarit	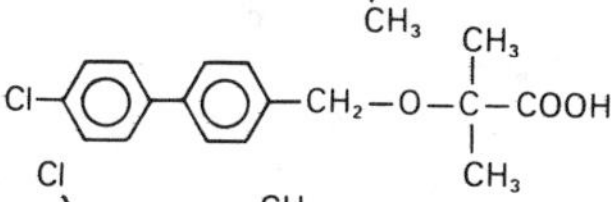 Cl-C_6H_4-C_6H_4-CH_2-O-$C(CH_3)_2$-COOH	4.01
Silvex	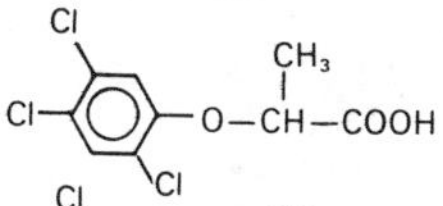 Cl$_3$-C_6H_2-O-CH(CH_3)-COOH	2.59
Dichloroprop	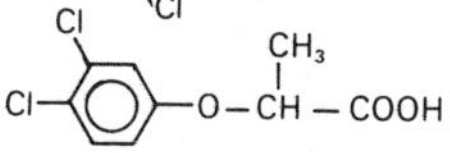 Cl$_2$-C_6H_3-O-CH(CH_3)-COOH	1.33
Clofibric acid	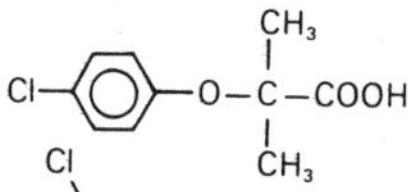 Cl-C_6H_4-O-$C(CH_3)_2$-COOH	1.00
2,4,5-T	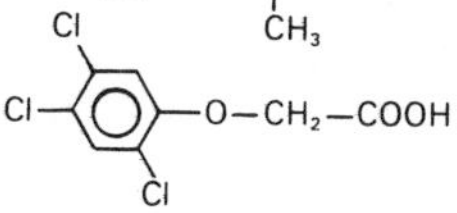 Cl$_3$-C_6H_2-O-CH_2-COOH	0.72
2,4-DB	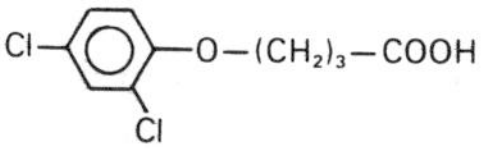 Cl$_2$-C_6H_3-O-$(CH_2)_3$-COOH	0.43
2,4-D	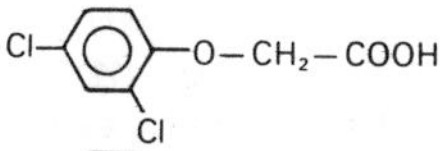 Cl$_2$-C_6H_3-O-CH_2-COOH	0.26
MCPA	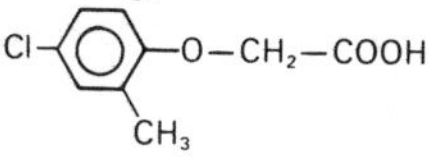 Cl(CH_3)-C_6H_3-O-CH_2-COOH	0.22

Fig. 2. Relative Peroxisome Proliferative Activity of Chlorinated Phenoxyacetic Acids and Related Compounds in Cultured Rat Hepatocytes. (*R.P. = Relative potency for induction of palmitoyl CoA oxidation. Data summarized from Lewis et al. [20])

potential carcinogenicity and whether the carcinogenicity
of chlorinated phenoxyacetic acid derivatives can be
totally accounted for by their peroxisome proliferative
activity. In this respect, it is interesting to point out
that trichloroacetic acid, a relatively weak peroxisome
proliferator   (20),   has   been   shown   to   be
hepatocarcinogenic (see ref. 5). At least in the case of
trichloroacetic   acid,   it   appears   that   peroxisome
proliferation alone cannot completely account for its
carcinogenic activity.

Scheme for Hazard Identification

Because of the complexity of SAR analysis a formalized
scheme is used for detecting potential carcinogens, so
as to ensure that all factors are properly considered.
One such scheme is shown in Fig. 3. After consideration
of the physicochemical properties, exposure pattern,
degradation potential, metabolism and pharmacokinetics,
three categories of criteria should be used for
suspecting chemical compounds of carcinogenic activity.
These are: (1) structural criteria, (2) functional
criteria, and (3) "guilt-by-association" criterion.

(1) Structural criteria are based on SAR analysis. Two
basic approaches may be used: firstly, structural
analogies with established types of chemical carcinogens,
and, secondly, considerations of molecular size, shape
and symmetry, and of electron distribution and steric
factors in and around functional groups, independently
from any possible analogy with other compounds. Some of
the structural criteria (see Appendix V of ref. 5 for
details) are listed below:

1. The presence of an amino, dimethylamino,
nitroso or nitro group directly linked to an
aromatic conjugated double-bonded system,
especially if the amine or amine-generating
group is at the terminal end of the longest
conjugated double-bond arrangement of the
molecule.

2. Polycyclic structure with three to six
rings that mimic the angular ring distribution
of carcinogenic polycyclic hydrocarbons. The
likelihood of carcinogenicity increases if the
molecule is asymmetric, contains bay-region
benzo ring, blocked L-region, and free peri
position adjacent to the bay-region benzo
ring.

3. N-Nitroso, hydrazo, aliphatic azo, or
aliphatic azoxy structures and all 1-aryl-3,3-

dialkyltriazenes and 1,1-diaryl-2-acetylenic carbamates.

4. The presence of a sterically strained ring (e.g., epoxide, aziridine, lactones and sultones) in any type of structure. The likelihood of carcinogenicity may increase if the compound contains two or more of these reactive ring structures.

5. Any structural type of alkylating, arylating, or acylating agent or larger molecular assemblies incorporating such agents as chemically reactive moieties. The likelihood of carcinogenicity may increase if (i) the reactive intermediate generated can be stabilized by resonance or by inductive or mesomeric effects, (ii) the compound contains two or more of these reactive moieties, and/or (iii) the compound has a favorable molecular shape and size for DNA intercalation or is a structural analog of normal cellular constituents.

6. Low molecular weight carbamate, thio-carbamate, and thiourea derivatives.

7. The presence of a haloalkyl (particularly 1,2-dihalo), haloalkenyl (both vinylic and allylic), α-haloether, α-haloalkanol, or α-halocarbonyl grouping.

8. Low molecular weight aliphatic structures containing conjugated double bonds or isolated double bonds situated at the terminal end of an aliphatic chain.

9. Low molecular weight aldehydes. The likelihood of carcinogenicity increases if (i) the compound contains two terminal aldehyde groups separated by less than four interconnecting atoms, and/or (ii) presence of α,β-double bond.

10. Benzylic, allylic or pyrrolic esters if the acyloxy moiety is a good leaving group.

(2) <u>Functional criteria</u> represent the sum of pharmacological and toxicological capabilities which, irrespective of structural type, have been correlated with carcinogenic activity. Functional criteria are used in a complementary manner to structural criteria because structural considerations alone cannot forecast entirely new structural types of carcinogens. Some of the

functional criteria are:

1. agents that bring about _in vitro_ cell transformation

2. oncogene activators

3. mutagens and clastogens (chromosomal aberration)

4. aneuploidogens/spindle poisons

5. teratogens and inhibitors of spermatogenesis

6. agents showing electrophilicity and/or covalent binding to macromolecules

7. peroxisome proliferators and/or reactive oxygen generators

8. inhibitors of intercellular communication

9. strong inducers of cell proliferation and/or of mixed-function oxidase(s)

10. antineoplastic agents

11. immunosuppressive agents

12. agents causing hormonal imbalance

13. strong surface active agents, H-bond reactors, or chelators

14. persistent cytotoxic agents

15. agents causing rough endoplasmic reticulum degranulation (detachment of ribosomes)

16. certain inhibitors of mitochondrial respiration

At least some of these criteria are so highly correlated with carcinogenicity that they have been developed as short-term tests. One good approach to improve the use of the functional criteria is to carry out a systematic chemical class-by-chemical class study of the correlation between the particular functional criterion in question and carcinogenicity. Not all chemical classes are expected to demonstrate a certain functional criterion endpoint. The class-specific approach should enhance the usefulness of the functional criteria approach in predicting potential carcinogenicity.

(3) The "Guilt-by association" criterion points to the possible carcinogenic potentiality of compounds which, although found inactive under some "standard" conditions of animal bioassay (and/or mutagenicity testing), belong in chemical classes in which several other compounds were found to be potent and multi-target carcinogens, for example the 5-nitrofuran type urinary antibacterials. These compounds should be reevaluated to determine if retesting under more strigent conditions is warranted.

Brief Overview of Structural Features of Selected Classes of Carcinogenic Pesticides

1. **Organophosphorus pesticides:** For various reasons, organophosphorus pesticides represent a very difficult group for the evaluation of potential carcinogenicity based on SAR analysis. Table II summarizes the results of U.S. National Cancer Institute/National Toxicology Program (NCI/NTP) bioassays of 13 organophosphorus pesticides. Of these, two (tetrachlorvinphos and, by gavage, dichlorvos) gave positive results while four or five yielded equivocal results. Industry data submitted to EPA indicate that at least a number of other organophosphorus pesticides are/may be carcinogenic (22). Overall, virtually all these positive or equivocal organophosphorus pesticides are methyl or ethyl esters of phosphoric or thiophosphoric acids. Most of these pesticides are electrophilic and/or mutagenic but the correlation between carcinogenicity and electrophilicity/mutagenicity seems to be poor.

At least five critical factors affect the potential carcinogenicity of organophosphorus pesticides (see Fig. 4). The alkyl groups (R) are generally believed to be a major contributor to the genotoxicity of organophosphorus compounds although the role of the electrophilic phosphoryl moiety cannot be excluded. In general, the alkylating activity of organophosphorus compounds decreases with an increase of the size of the R group. Alkyl esters of thiophosphoric acids are not as effective alkylators as the phosphates. They can, however, be metabolically converted to phosphates by oxidative desulfuration or by thiono-thiolo rearrangement. The alkylating activity of organophosphorus compounds can be increased by increasing the electron-withdrawing capability of the leaving group. However, the same parameter can also make the phosphoryl moiety more susceptible to alkaline hydrolysis. Once the leaving group departs, the resulting dialkyl phosphate is no longer an alkylating agent.

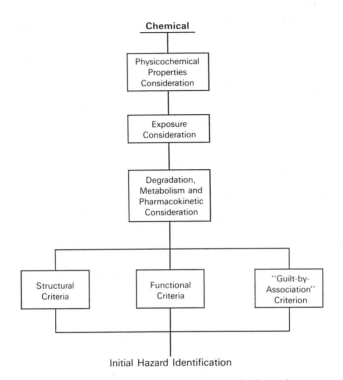

Initial Hazard Identification

Fig. 3. Scheme for Hazard Identification Based on Structural, Functional and "Guilt-by-association" Criteria.

Organophosphorous Pesticides

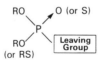

Critical Factors:

1. Nature of R group: size, branching, halogenation.

2. Phosphate versus thiophosphate.

3. Electron - withdrawing capability of the leaving group.

4. Detoxification by hydrolysis or -SH compounds.

5. Breakdown products.

Fig. 4. Critical Factors Affecting Carcinogenic Activity of Organophosphorus Pesticides.

TABLE II

Organophosphorus Pesticides That Have Been Tested by U.S. National Cancer Institute/National Toxicology Program

pesticide	tr	salm	route	mr	fr	mm	fm
AZINPHOSMETHYL	069	+w	FEED	E	N	N	N
COUMAPHOS	096	–	FEED	N	N	N	N
DIAZINON	137	–	FEED	N	N	N	N
DICHLORVOS	010	+	FEED	N	N	N	N
DICHLORVOS	342	+	GAV	SE	EE	SE	CE
DIMETHOATE	004	++	FEED	N	N	N	N
DIOXATHION	125	+	FEED	N	N	N	N
FENTHION	103	–+w	FEED	N	N	E	N
MALAOXON	135	–	FEED	N	N	N	N
MALATHION	024	–	FEED	N	N	N	N
MALATHION	192	–	FEED	N	N		
METHYL PARATHION	157	+	FEED	N	N	N	N
PARATHION	070	+w–	FEED	E	E	N	N
PHOSPHAMIDON	016	+	FEED	E	E	N	N
TETRACHLORVINPHOS	033	–	FEED	N	P	P	P

Code: tr = NCI or NTP technical report numbers; salm = Ames test (+, positive; +w, weakly positive; –, negative; ONT, on test); mr = male rats; fr = female rats; mm = male mice; fm = female mice; evidence for carcinogenicity (CE, clear evidence; P, positive; SE, some evidence; E or EE, equivocal evidence; IS, inadequate study; N or NE, negative).

Detoxification of organophosphorus pesticides before they can reach their target sites is probably the main reason for poor correlation between carcinogenicity and electrophilicity/mutagenicity. The problem is further complicated by the fact that several different enzymes are involved in the metabolic detoxification of organophosphorus pesticides. For example, paraoxon, tetrachlorvinphos and dimethoate are preferentially detoxified by A-esterase (paraoxonase), GSH-dependent S-alkyltransferase and carboxyesterase (aliesterase), respectively whereas chlorfenvinphos is mainly detoxified by NADPH-dependent oxidative dealkylation (23). Furthermore, significant species differences have been observed in the rate of detoxification; for example, the relative rates of dealkylation of chlorfenvinphos by rat, mouse, rabbit and dog livers are 1, 8, 24 and 88, respectively (23). A systematic study of detoxification pattern of organophosphorus pesticides is critically needed before we can obtain more accurate prediction from SAR analysis.

Terminal or vicinal halogenation of the R group can be expected to increase the electrophilic reactivity of the breakdown product(s). Besides the alkyl group, other breakdown products of organophosphorus pesticides may also contribute to carcinogenicity. For example, it has been suggested (24) that the leaving group of tetrachlorvinphos (Fig. 5) can contribute to carcinogenicity by generating the electrophilic intermediate, chloromethyl 2,4,5-trichlorophenyl ketone.

2. Carbamate/thiocarbamate pesticides: Considerably less information on carbamate pesticides is available in the open literature for establishing a clear SAR pattern. However, from studies on urethan and its analogs (3,25), some structural features that favor carcinogenicity/ mutagenicity can be discerned (see Fig. 6). They are: (i) a small alkyl group at the carboxy end. Vinyl carbamate is by far the most potent carbamate carcinogen. Urethan (ethyl carbamate) has been postulated to be metabolically activated by dehydrogenation to vinyl carbamate. However, the demonstration of carcinogenicity of methyl carbamate by NTP (Technical Report No. 328) suggests that dehydrogenation is not an obligatory route for carbamate activation. Two carbamate pesticides (asulam, benomyl) reported to be carcinogenic (22) are N-substituted derivatives of methyl carbamate. (ii) Some fourteen 1,1-diaryl-2-acetylenic carbamates of the structure given in Fig. 6 have been shown to be carcinogenic. The aryl and acetylenic moieties are probably involved in stabilizing the carbonium ion which would arise after departure of the carbamoyloxy moiety. (iii) Replacement of the alkoxy group by chlorine yields a potent carcinogen,

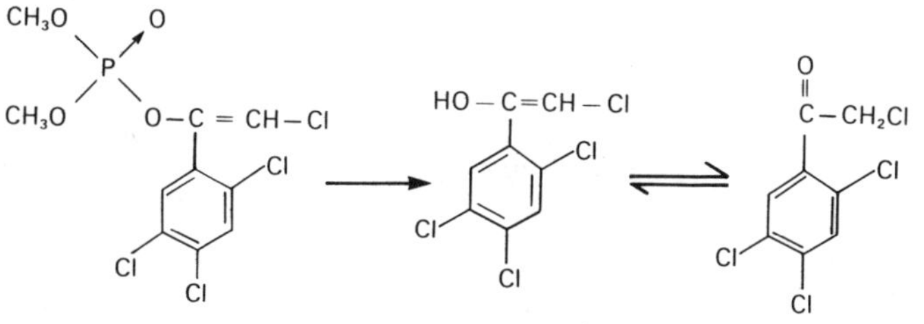

Tetrachlorvinphos

chloromethyl
2,4,5-trichlorophenyl
ketone

Fig. 5. Leaving Group as Potential Electrophilic Carcinogenic Intermediate of the Organophosphorus Pesticide, Tetrachlorvinphos.

CARBAMATES

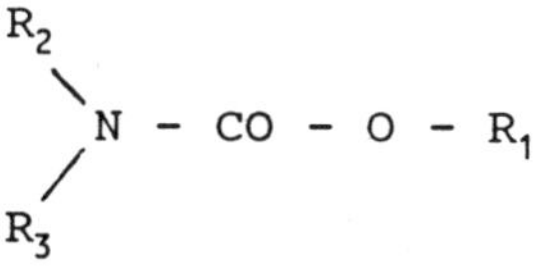

$$\begin{array}{c} R_2 \\ \diagdown \\ \diagup \\ R_3 \end{array} N - CO - O - R_1$$

Salient structural features known to contribute to or affect carcinogenicity/mutagenicity:

1. Carcinogenic if R_2, R_3 = H; R_1 = vinyl, ethyl or methyl.

2. Carcinogenic if R_2, R_3 = H or alkyl; R_1 =

$$-\overset{\displaystyle Aryl}{\underset{\displaystyle Aryl}{\overset{|}{\underset{|}{C}}}} - C \equiv C - H$$

   (or R)

3. Carcinogenic if R_2, R_3 = CH_3; $- OR_1$ replaced by $- Cl$.

4. N,N-Disubstitution with bulky group decreases carcinogenicity/mutagenicity.

5. Mutagenic if either R_2 or R_3 is acyloxy.

Fig. 6. Salient Structural Features Known to Contribute to or Affect the Carcinogenicity and/or Mutagenicity of Carbamates.

dimethylcarbamyl chloride. (iv) With the exception of good leaving groups, N,N-disubstitution generally decreases carcinogenicity/mutagenicity probably by decreasing the possibility of metabolic activation of the amino end. (v) N-Substitution with a good leaving group such as an acyloxy group can yield a potent direct-acting mutagen. These results suggest that there are three potential electrophilic sites in carbamates: the alkyl group, the carbamoyl group and the activated amino end. These structural features should be useful in evaluating the potential carcinogenicity of carbamate pesticides.

A carcinogenicity concern of another type associated with carbamate pesticides is their relative ease of N-nitrosation. Nitrosating agents can be readily found in the environment (e.g., nitrite in saliva) and the acidic condition of the stomach favors the N-nitrosation reaction. Once nitrosated, the resulting N-nitroso-carbamate has a very high probability of being carcinogenic. Bioassay studies on N-nitroso compounds indicate that, of the close to 400 N-nitroso compounds that have been tested, some 80-90% are carcinogenic. The significance of possible _in vivo_ formation of N-nitroso compounds from carbamate pesticides requires more extensive consideration.

A number of thiocarbamate and dithiocarbamate pesticides have been shown to be carcinogenic or mutagenic (3, 25). These include diallate, triallate and sulfallate. Their common structural feature appears to be the presence of the S-chloroallyl moiety that can yield chloroacrolein as the most probable ultimate carcinogen/mutagen. A number of metal salts of dialkyldithiocarbamate (particularly dimethyldithiocarbamates) and the closely related tetramethylthiram and tetramethylthiram monosulfide have been shown to be mutagenic in the Ames test (25). Several of these compounds were positive in the NCI screening test (3, 25) but thus far only ziram (NTP Technical Report No. 238) was found positive when retested in a full-fledged bioassay, inducing thyroid tumors in male rats. Bisethylenedithiocarbamates as a class are suspect carcinogens because they can all yield ethylenethiourea, a proven carcinogen, after metabolism (25).

3. _Organohalogen pesticides:_ A large number of organohalogen pesticides have been tested for carcinogenicity. Table III shows that, among the 28 organohalogen pesticides tested by U.S. National Cancer Institute/National Toxicology Program (NCI/NTP), 17 are positive in at least one animal species. Based on chemical structure, they can be classified as halogenated alkanes, alkenes, cycloalkanes,

TABLE III

Organohalogen Pesticides That Have Been Tested by U.S.
National Cancer Institute/National Toxicology Program

pesticide	tr	salm	route	mr	fr	mm	fm
ALDRIN	021	ONT	FEED	E	E	P	N
CHLORDANE, TECHNICAL	008		FEED	N	N	P	P
CHLOROBENZILATE	075	−	FEED	E	E	P	P
3-CHLORO-2-METHYLPROPENE	300	+w	GAV	CE	CE	CE	CE
CHLOROTHALONIL	041	−	FEED	P	P	N	N
p,p'-DDE	131	−	FEED	N	N	P	P
DDT	131	−	FEED	N	N	N	N
1,2-DIBROMO-3-CHLOROPROPANE	206	+	INHAL	P	P	P	P
1,2-DIBROMO-3-CHLOROPROPANE	028	+	GAV	P	P	P	P
1,2-DIBROMOETHANE	086	ONT	GAV	P	P	P	P
1,2-DIBROMOETHANE	210	ONT	INHAL	P	P	P	P
1,2-DICHLOROBENZENE	255	−	GAV	N	N	N	N
1,4-DICHLOROBENZENE	319	−	GAV	CE	NE	CE	CE
1,2-DICHLOROPROPANE	263	+w	GAV	NE	EE	SE	SE
1,3-DICHLOROPROPENE	269	+	GAV	CE	SE	IS	CE
(TELONE II)							
DICOFOL	090	−	FEED	N	N	P	N
DIELDRIN	021	−	FEED	N	N	E	N
DIELDRIN	022	−	FEED	N	N		
ENDOSULFAN	062	−	FEED	I	N	I	N
ENDRIN	012	−	FEED	N	N	N	N
p,p'-ETHYL-DDD (PERTHANE)	156	+	FEED	N	N	N	E
HEPTACHLOR	009	−	FEED	N	N	P	P
LINDANE	014	−	FEED	N	N	N	N
METHOXYCHLOR	035	−	FEED	N	N	N	N
MIREX	313	−	FEED	CE	CE		
PENTACHLORONITROBENZENE	061	−	FEED	N	N	N	N
PENTACHLORONITROBENZENE	325	−	FEED			NE	NE
PENTACHLOROPHENOL, DOWICIDE	349		FEED			CE	CE
PENTACHLOROPHENOL, TECHNICAL	349	−	FEED			CE	SE
PHOTODIELDRIN	017	+	FEED	N	N	N	N
p,p'-TDE	131	−	FEED	E	N	N	N
TOXAPHENE	037	+	FEED	E	E	P	P
2,4,6-TRICHLOROPHENOL	155	−	FEED	P	N	P	P

Code: tr = NCI or NTP technical report numbers; salm =
Ames test (+, positive; +w, weakly positive; −, negative;
ONT, on test); mr = male rats; fr = female rats; mm =
male mice; fm = female mice; evidence for carcinogenicity
(CE, clear evidence; P, positive; SE, some evidence; E or
EE, equivocal evidence; IS, inadequate study; N or NE,
negative).

cycloalkenes, aromatics, bridged aromatics, phenols and phenoxyacetic acids. They can act by a variety of mechanisms.

In general, lower halogenated alkanes and alkenes tend to be genotoxic carcinogens (see ref. 4). They can be direct-acting as well as indirect-acting. The structural features that are predictive of potent carcinogens are: (i) presence of active halogen(s) in terminal position(s), (ii) presence of additional electrophilic functional group(s) (e.g., another halogen, double bond that can be epoxidized), (iii) vicinal dihalogenation (e.g., 1,2-dihaloethane) which allows an additional mode of activation by GSH conjugation to yield reactive episulfonium ions, and (iv) favorable physicochemical properties such as high volatility, low molecular weight, etc.

Halogenated aromatics and polyhalogenated hydrocarbons tend to be epigenetic carcinogens (see ref. 4). Most of these are carcinogenic partly because of persistence in the body, so that structural features that increase metabolism can be expected to decrease the carcinogenic potential. For example, replacement of chlorine atoms at p,p'-positions of DDT with the easily metabolizable methoxy groups yields a noncarcinogenic compound, methoxychlor. Other structural features that may increase metabolism include ring para positions that are unsubstituted and adjacent unsubstituted ring positions.

4.   Other miscellaneous classes; Besides the major classes of carcinogenic pesticides summarized above, specific structural features may account for carcinogenicity of some pesticides. For example, the amidine pesticide, chlordimeform, is carcinogenic most probably

Chlordimeform

because of its degradation products/metabolites 4-chloro-o-toluidine (positive in a NCI bioassay) and/or 3-(2-methyl-4-chlorophenyl)-1,1-dimethylurea (an analog of monuron, also positive in a NTP bioassay). It is interesting to note that chlordimeform itself is not mutagenic in several short-term tests but its metabolite/degradation product, 4-chloro-o-toluidine is positive in the Ames test and a mammalian spot test (27). Based on chemical properties alone, the acid/metabolic lability of the amidine structure should not be unexpected. All amidine pesticides should be evaluated for the carcinogenic potential of their degradation products/metabolites.

Captan and folpet, two phthalimide fungicides, have been shown or reported to be carcinogenic (see refs. 5, 22) and mutagenic (see ref. 28). While the mechanism of

Captan   Folpet

their action is not clear, the carcinogenic/mutagenic action of these compounds is most likely associated with the N-S-CCl$_3$ moiety, a direct-acting electrophile which is not effectively detoxified by noncritical cellular nucleophiles. Cellular thiols (RSH) are expected to react with (the electrophilic) sulfur in the N-S-CCl$_3$ moiety. However, the disulfide (R-S-S-CCl$_3$) metabolite thus formed may rearrange and break down to yield

thiophosgene (Cl-CS-Cl), another electrophilic intermediate with carcinogenic potential. Another closely related carcinogenic fungicide, captafol (an analog of captan, with N-S-CCl$_2$-CHCl$_2$) is also expected to undergo the same type of reaction to yield Cl-CS-CHCl$_2$ as a possible electrophilic intermediate.

Alachlor and metolachlor, two closely related pre-emergence α-chloroacetanilide herbicides, have been reported to be carcinogenic (see ref. 5). Their

Alachlor   Metolachlor

carcinogenic activity is most likely related to the
active chlorine (in the α-chloroketo moiety) which can
depart to yield an alkylating intermediate.
Alternatively, the possible degradation products, o-
substituted anilines, are also suspect carcinogens.

A number of other structural types of pesticides
are/may be also of carcinogenicity concern. These
include the biphenyl ethers, hydrazides, substituted
phenylureas and substituted amino-s-triazines as
represented by nitrofen (NCI Technical Report No. 26),
daminozide (NCI Technical Report No. 83), monuron (NTP
Technical Report No. 266) and propazine (see ref. 29),

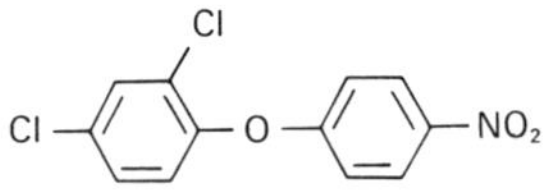

Nitrofen

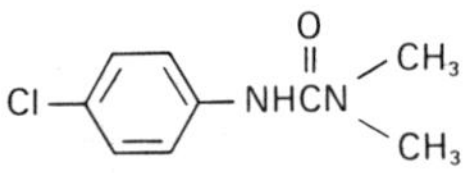

Monuron

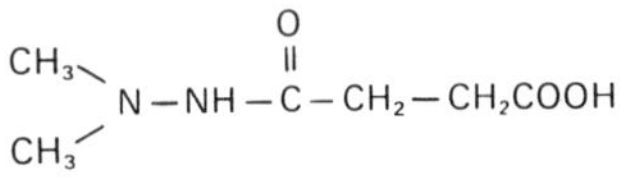

Daminozide

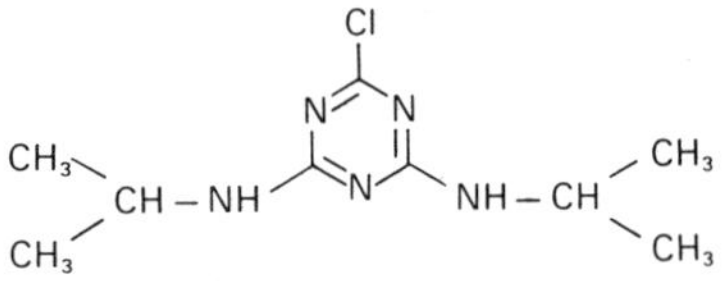

Propazine

respectively. More SAR information is needed to clearly
delineate the structural features in each of these
classes. For example, although nitrofen is mutagenic in
the Ames test (see ref. 30), its close analog, 1'-
(carboethoxy)ethyl 5-(2-chloro-4-[trifluoromethyl]-
phenoxy)-2-nitrobenzoate (lactofen, which also contains
the potentially electrophilic aromatic nitro group),
appears to be nongenotoxic and to exert its carcinogenic
action by inducing peroxisome proliferation (31).
Hydrazides as a class are relatively weak or marginally
active carcinogens (see ref. 3); the ease of hydrolytic
release of hydrazine compounds (more carcinogenic) from
hydrazides may be an important factor.

Concluding Comment

SAR analysis is a complex process, and is
meaningful only after consideration of all factors. With

sufficient supportive data SAR analysis is a useful tool for detecting carcinogens among new and existing pesticides. SAR should be used as a guiding principle in prioritizing testing for carcinogenicity as well as in choosing effective batteries of tests of toxicological endpoints relevant to carcinogenesis.

Disclaimer: The conclusions reached and scientific views expressed in this chapter are solely those of the authors and do not necessarily represent the views or policies of the U.S. Environmental Protection Agency. Mention of trade-names, commercial products or organizations does not imply endorsement by the U.S. Government.

Literature Cited

1. Arcos, J.C.; Argus, M.F. Chemical Induction of Cancer, Volume IIA (Polynuclear Compounds), Academic Press, New York, 1974.
2. Arcos, J.C.; Argus, M.F. Chemical Induction of Cancer, Volume IIB (Aromatic Amines and Azo Dyes), Academic Press, New York, 1974.
3. Arcos, J.C.; Woo, Y.-T.; Argus, M.F. (with the collaboration of Lai, D.Y.) Chemical Induction of Cancer, Volume IIIA (Aliphatic Carcinogens), Academic Press, Orlando, Florida, 1982.
4. Woo, Y.-T.; Lai, D.Y.; Arcos, J.C.; Argus, M.F. Chemical Induction of Cancer, Volume IIIB (Aliphatic and Polyhalogenated Carcinogens), Academic Press, Orlando, Florida, 1985.
5. Woo, Y.-T.; Lai, D.Y.; Arcos, J.C.; Argus, M.F. Chemical Induction of Cancer, Volume IIIC (Natural, Metal, Fiber and Macromolecular Carcinogens), Academic Press, San Diego, California, 1988.
6. Woo, Y.-T.; Arcos, J.C.; Lai, D.Y. In Handbook of Carcinogen Testing, Milman, H.A.; Weisburger, E.K., Eds.; Noyes Publ., Park Ridge, New Jersey, 1985, pp. 2-25.
7. Woo, Y.-T.; Arcos, J.C.; Lai, D.Y. In Chemical Carcinogens: Activation Mechanisms, Steric and Electronic Factors and Reactivity, Politzer, P.; Martin, Jr., F.J., Eds.; Elsevier, Amsterdam, 1988, pp. 1-31.
8. Sugimura, T.; Sato, S.; Wakabayashi, K. In Chemical Induction of Cancer, Volume IIIC (Natural, Metal, Fiber and Macromolecular Carcinogens), Woo, Y.-T.; Lai, D.Y.; Arcos, J.C.; Argus, M.F., Academic Press, San Diego, California, 1988, p.681.
9. Dipple, A.; Moschel, R.C.; Bigger, C.A.H. In Chemical Carcinogens, 2nd edn., Searle, C.E., Ed.;

ACS Monog. No. 182, American Chemical Society, Washington, DC, 1984, pp.41-163.

10. Guengerich, F.P. Cancer Res. 1988, 48, 2946-2954.

11. Reed, G.A. Environ. Carcino. Revs. (J. Environ. Sci. Hlth.) 1988, C6, 223-259.

12a. Krauss, R.S.; Eling, T.E. Carcinogenesis 1987, 8, 659-664.

12b. Anders, M.W; Dekant, W.; Elfarra, A.A.; Dohn, D.R. CRC Crit Rev.Toxicol. 1988, 18, 311-341.

13. Barbacid, M. Carcinogenesis 1986, 7, 1037-1042.

14. Poland, A.; Knutson, J.L. Ann. Rev. Pharmacol. Toxicol. 1982, 22, 517-554.

15. Thakker, D.R; Yagi, H.; Nordqvist, M.; Lehr, R.E.; Levin, W.; Wood, A.W.; Chang, R.L.; Conney, A.H.; Jerina, D.M. In Chemical Induction of Cancer, Volume IIIA (Aliphatic Carcinogens), Arcos, J.C.; Woo, Y.-T.; Argus, M.F., Academic Press, Orlando, Florida, 1982, pp. 727-747.

16. Wislocki, P.G.; Chang, R.L.; Wood, A.W.; Levin, W.; Yagi, H.; Hernandez, O.; Mah, H.D.; Dansette, P.M.; Jerina, D.M. Cancer Res. 1977, 37, 2608-2611.

17. White, Jr., K.L. Environ. Carcino. Revs. (J. Environ. Sci. Hlth.) 1986, C4, 163-202.

18. Pearson, R.G; Songstad, J. J. Amer. Chem. Soc. 1967, 89, 1827-1836.

19. Reddy, J.K.; Lalwani, N.D. CRC Crit Rev. Toxicol. 1983, 12, 1-58.

20. Lewis, D.F.V.; Lake, B.G.; Gray, T.J.B.; Gangolli, S.D. Arch. Toxicol. 1987, suppl. 11, 39-41.

21. Rao, M.S.; Lalwani, N.D.; Watanabe, T.K.; Reddy, J.K. Cancer Res. 1984, 44, 1072.

22. Pesticide & Toxic Chemical News, 1985, 13(48), 14.

23. Murphy, S.D. In Casarett and Doull's Toxicology, 3rd edn.; Klaassen, C.D.; Amdur, M.O.; Doull, J., eds.; Macmillan, New York, 1986, p. 534.

24. Burchfield, H.P.; Storrs, E.E. In Environmental Cancer Kraybill, H.F.; Mehlman, M.A., Eds.; Wiley, New York, 1977, pp. 319-371.

25. Woo, Y.-T. Environ. Carcino. Revs. (J. Environ. Sci. Hlth.) 1983, C1, 97-133.

26. Aizawa, H. Metabolic Maps of Pesticides, Academic Press, Orlando, Florida, 1982.

27. Lang, R. Mutat. Res. 1984, 135, 219-224.

28. Garrett, N.E.; Stack, H.F.; Waters, M.D. Mutat. Res. 1986, 168, 301-325.

29. Pesticides & Toxic Chemical News, 1988, 16(15), 46.

30. Zeiger, E. Cancer Res. 1987, 47, 1287-1296.

31. Butler, E.G.; Tanaka, T.; Ichida, T.; Maruyama, H.; Leber, A.P.; Williams, G.M. Toxicol. Appl. Pharmacol. 1988, 93, 72-80.

RECEIVED July 18, 1989

Chapter 12

Sequential Assessment of Carcinogenic Hazard and Risk

John Ashby

ICI Central Toxicology Laboratory, Alderley Park, Macclesfield, Cheshire, England

As for any other chemical, knowledge of the rodent carcinogenicity of a pesticide usually provides the most important piece of information for estimating its possible carcinogenic hazard to humans. In the case of a rodent carcinogen, the estimation of possible human hazard can be refined by considering the probable mechanism of carcinogenic action in rodents. This can be done using chemical, genetic, biochemical and metabolic data in conjunction with consideration of the extent and type of the carcinogenic response observed in rodents. Some rodent carcinogens will thereby be classed as presenting a possible human carcinogenic hazard while others will not. If a possible human hazard is recognized, the separate process of estimating the carcinogenic risk to humans can proceed. Among other factors, this will be influenced by the nature, level and duration of human exposure. Given that several separate steps are required to estimate a carcinogenic risk to humans, it is concluded that the unqualified phrase a "carcinogenic pesticide" is of little practical value.

Two decades or so ago, carcinogens were a rarity. To anyone with but a passing interest in the subject the major carcinogens were well known. Almost without exception they were regarded as laboratory curiosities; they did not seem to be represented in normal life. Thus, with good faith and a reasonable amount of circumstantial evidence, it was usually assumed that drugs, food additives, pesticides and industrial chemicals were non-carcinogenic. The suspicion existed that there might have been further instances of chemically-induced carcinogenesis among workers in the chemical industry, but even there the trend to improve industrial hygiene reduced the chance of this happening.

Major attention began to be focused on the subject of chemical carcinogenicity in the early 1970's. The discussion centered on the prospect that if chemical usage continued to grow at contemporary rates, then an increase in the incidence of human cancers could be expected to occur in the late 1980's. This implicitly endorsed the safety of those chemicals in use at that time; the danger was perceived to reside with new chemicals and products.

During the past twenty years, two major trends of relevance to this symposium have become evident. First, the number of identified rodent carcinogens has increased more than could have been anticipated, almost to the point that non-carcinogens have now become the rarity. Second, the projected epidemic of chemically-induced cancers in man has not come to pass. Given the apparent conflict between these two trends, a

0097–6156/89/0414–0201$06.00/0

symposium on the carcinogenicity of pesticides would appear to be esoteric to the scientist, although the public continues to be deeply concerned about this issue. In fact, the conflict of the above two trends has not proven as damaging as it might. Two deeper realizations have occurred over the same period, and these continue to strain our ability to derive a balanced view of the subject. The first is that detailed bioassays of chemicals for carcinogenicity in rodents over the past decade have revealed subtle carcinogenic effects for many chemicals that otherwise give no indications of carcinogenicity. The extent to which some of these carcinogenic effects describe a property of the test species rather than of the test chemical is now being widely discussed. Second, great progress has recently been made in the science of risk estimation in chemical carcinogenicity. Thus, the primary point of interest is no longer whether a chemical is a rodent carcinogen, but rather, whether it is likely to present a carcinogenic risk in a given human exposure situation. The reconciliation of these two concepts presents a subtle challenge to toxicologists.

The Perception of Carcinogenic Risk

The seminal changes that have taken place in our perceptions of carcinogenic hazard and risk over the past few years can be illustrated as follows. It is now generally accepted that attention should be given to the question of whether a given organic chemical might present a human carcinogenic hazard. Quantitation of that hazard might indicate an unacceptable risk. However, it is now also accepted that even an established human carcinogen can be handled by humans without a significant carcinogenic risk so long as adequate protective measures are taken, i.e., so long as the levels of exposure are sufficiently low. Equally, a body of evidence is growing that indicates exposure to some rodent carcinogens is unlikely to present any carcinogenic hazard to humans, regardless of exposure.

The guiding principles in deciding on the regulatory course of action with respect to a suspect carcinogen must be based on an understanding of the mechanism of carcinogenic action of the compound in question. It is no longer useful to refer to the carcinogenicity of, for example, sweeteners or of pesticides in isolation. Rather, this must be qualified by some knowledge or estimation of the mechanism of carcinogenic action and the projected extent of exposure to the agent. Relating this to the carcinogenicity of pesticides yields the following primary categories of concern.

a)   Some pesticides will prove non-carcinogenic to both rodents and humans under any conditions of exposure. They may, of course, possess other toxicities that could present a possible human hazard, but that is not pursued herein.

b)   Some pesticides will prove carcinogenic to rodents, yet it will be possible to predict that no human carcinogenic hazard is likely to be present during their manufacture, use, or ingestion of residues in food.

c)   Some pesticides will prove carcinogenic to rodents, and by means of the consideration discussed later herein, it may be possible also to anticipate a human carcinogenic hazard from them. The risk associated with this hazard will be magnified or reduced dependent upon conditions of exposure. This suggests possibly different risks for the manufacture and dispensing of such agents, as compared to their ingestion by humans as residues.

d)   Some pesticides will prove to be human carcinogens and must be regulated accordingly.

There are many hundreds of pesticides in use worldwide. For a minority of these there are adequate experimental data to enable their classification as either non-carcinogens or animal carcinogens. It is tempting to relate the remainder of this paper to a discussion of these agents, but this would be idly repetitive of current initiatives being taken on these agents by the major regulatory authorities. Rather, attention has been focused on the principles that are of importance when considering agents for which appropriate toxicological data do not exist. These principles are independent of the end-use of a chemical, so they are as equally applicable to fire-retardants as they are to pesticides. Risk estimations must utilize toxicological principles and data, where they exist, along with exposure levels experienced through the end-use of the compound.

Sequential Assessment of Carcinogenic Hazard

A range of studies is usually required before a confidant assessment of carcinogenic hazard to humans can be derived. If a possible human hazard is defined, a separate set of studies may be required in order to quantitate the risk associated with the hazard. The risk will vary with usage pattern, and some perception of acceptable risk will usually be required. The advantage of adhering to a sequence in the assessment of carcinogenic hazard and risk is that it can provide indications of where resources should be allocated. The alternative is to ignore major hazards while refining perceptions of a negligible or non-existent hazard associated with a pesticide.

In order to illustrate clearly the intended message of this paper the development of four model organic chemicals as possible pesticides is charted in tabular form. The hypothetical conclusions at each stage of evaluation are realistic and representative, albeit the specific results suggested might be different were these chemicals actually to be prepared and tested as described.

Four hypothetical organic chemicals based on the benzopyran nucleus have been selected for consideration. This nucleus has neither been associated with genotoxicity or carcinogenicity, nor are such properties expected. Therefore, the possible activity of substituents becomes the potentially important issue. The reader is asked to assume that compounds I-IV possess pesticidal properties, and that they are also carcinogenic to rodents; they are, therefore, "carcinogenic pesticides."

The point of Table I is to illustrate how this epithet, "carcinogenic pesticides," can be qualified such that four quite distinct conclusions regarding carcinogenic hazard exist, with only one requiring regulatory action to prevent human exposure. The text in the table is contrived to illustrate principles, but it has been restrained so as to remain realistic. This table is entirely theoretical, but is representative of the types of assessments and conclusions which might be drawn from the data presented. The concepts implied are discussed in detail elsewhere (1-3).

The sequence indicated in the table on evaluation for possible carcinogenic hazard to man involves first the assessment of genotoxicity. This includes evaluation for possible sites of DNA reactivity, genotoxicity studies in vitro, and genotoxicity studies in vivo. These would be followed by what is considered more traditional toxicologic studies, including sub-acute studies and cancer bioassay in rodents by gavage. Finally, the mechanism of carcinogenic action must be postulated, based on all available evidence. Then, risk of carcinogenic hazard to humans can be assessed.

Compounds I and II represent chemicals for which there are data showing positive carcinogenicity in one highly sensitive rodent species, but these results probably uniquely relate to the physiology and biochemistry of these species rather than to the properties of the chemical itself. Thus, the conclusion for these two compounds is that risk estimation for humans is unnecessary. Compound III shows results indicating genotoxicity but the positive carcinogenicity results are such that they would probably not relate to the genotoxicity. Such a compound would need to be regulated in situations only where human exposure would be at elevated levels. Models for risk estimation of such compounds should accommodate the concept of thresholds. Compound IV is one in which all toxicologic evidence indicates positive carcinogenicity. Risk estimations need to be made and should be used in the regulation of such a material.

Table I. Hypothetical outcome of testing four hypothetical pesticides carcinogenic hazard

Stage of Evaluation for Possible Carcinogenic Hazard to Man	(I)	(II)
Evaluation for possible sites of DNA reactivity, based on historical precedents and chemical knowledge.	No sites expected to be electrophilic (DNA-reactive). Not expected that metabolism in mammals would generate such sites.	Similar comments as for (I). Hydrolytic opening of the anhydride expected, yielding the corresponding di-acid.
Genotoxicity *in vitro*. *Salmonella* (Ames) Clastogenicity (Lymph.) DNA damage (UDS) (3 tests selected that are acceptable by regulatory authorities)	- - -	- w+ - (w+ only at high doses: possible pH or osmolarity effect)
Genotoxicity *in vivo*. Bone marrow MN assay Liver UDS test (2 *in vivo* tests used consistent with some recent regulatory guidelines)	(-) (-) (Some investigators would not conduct *in vivo* experiments on this chemical, but results shown)	(-) (-) (Same comments as for I)
Sub-acute toxicity studies (28 day)	Non-toxic up to dose-levels of 5 g/kg, in mice or rats. Evidence of hyalin droplets in male rat kidney at higher dose-levels.	Non-toxic, but evidence of liver enlargement at dose-levels of 1 g/kg.
Carcinogenicity bioassay via oral gavage. (F344 rats and B6C3F1 mice, as conducted by the US NTP)	+ Some evidence (kidney) No evidence No evidence No evidence	+ No evidence No evidence Clear evidence (liver) Clear evidence (liver)
Consideration of mechanism of carcinogenic action, given the total database and the extent of current knowledge in these matters.	Male rat kidney tumors may be a male rat specific response. Genotoxicity unlikely to be implicated.	The mouse liver tumors may be uniquely related to the effect of the di-acid on fat metabolism in the mouse liver, and the subsequent mitotic activity.
Risk assessment for carcinogenic hazard to humans.	Probably no hazard, carcinogenic effect probably o rat-specific. Risk estimation unnecessary.	Probably no hazard, carcinogenic effect probably rodent specific. Risk estimation unnecessary.

for carcinogenic potential with a view to discerning the relative
to exposed humans

(III)	(IV)	Conclusions drawn
The aromatic -NHAc group could prove DNA-reactive after metabolism. The -CH₂OH group may oxidize to the corresponding acid (-COOH), thereby aiding excretion *in vivo*.	The aromatic -NO₂ groups could prove DNA-reactive after metabolism. The epoxide group expected to be DNA-reactive.	Compounds III and IV alerted as possible genotoxins and carcinogens.
+ (with liver S9) + (with liver S9) W+	+ (S9) + (-S9) W+ (Active in Ames test with NR⁻ strains suggesting epoxide also active)	Compounds III & IV confirmed as *in vitro* genotoxins. Weak clastogenicity of II may be artefact. I confirmed as non-genotoxic.
− − (second test justified because some arylamines are primarily active in the rodent liver)	+ (+) (second test would not normally be conducted, the agent being genotoxic *in vivo*)	Compounds I-III of low carcinogenic potential. Compound IV a confirmed potential carcinogen.
Effects on body weight gain at doses of 1.5 g/kg. Evidence of minimal thyroid gland hyperplasia at top dose-level.	Effects on body wt gain at doses of 500 mg/kg. Liver hypertrophy, & hyperplasia in non-glandular stomach. Depression of white blood count above 200 mg/kg.	Compound IV remains a potential genotoxic rodent carcinogen. Compounds I-III may induce non-genotoxic carcinogenic effects in rodents after lifetime exposure.
+ Some evidence (thyroid) Clear evidence (thyroid) No evidence No evidence	+ Clear evidence (stomach) Clear evidence (stomach) Clear evid.(liver,ovary) Clear evid.(liver,ovary)	Compounds I-IV are carcinogenic to rodents.
Many genotoxic and non-genotoxic arylamines are hyperplastic to the rat thyroid. The relevant carcinogenic effects seen are probably not related to genotoxicity (2).	The rat stomach tumors are probably due to the epoxide function. The mouse ovary/liver tumor may be related to the genotoxic -NO₂ subst.	Compounds I-III probably non-genotoxic rodent carcinogens, effects will probably not be applicable to man at the dose-levels likely to be encountered. Compound IV is a genotoxic rodent carcinogen, and as such, a possible human carcinogen.
The carcinogenic effects in rodents probably only relevant to man at elevated dose-levels. The *in vitro* genotox suggests that estimations of human exposure are worth deriving, leading to risk estimations using models that accommodate thresholds.	Assume carcinogenic hazard to exposed humans. Conduct appropriate risk estimations using available data and conservative models for extrapolation.	Distinct and graded risk estimations indicated. For two of the rodent carcinogens (I and II) probably no human hazard exists, at any exposure level, and therefore no risk.

Conclusions

Agrochemicals, including pesticides, are distinguished by the fact that they are purposely introduced into the environment in measured amounts. Further, information is often available concerning their residence time in the soil or on crops, including their average concentration in environmental products such as food and milk. Inevitably, these levels are very low, usually in the ppb-ppm range. If such an agent is defined as a rodent carcinogen, or if it is classified as a probable rodent carcinogen based on surrogate studies, three mutually dependent questions arise. First, is the agent likely also to present a carcinogenic hazard to humans? Second, if a human hazard is considered likely, what is the extent of the human carcinogenic risk at the levels of exposure anticipated? Third, is that risk commensurate with the advantages attendant to its use when considered alongside the more obvious risk factors in life? The fact of rodent carcinogenicity of a pesticide is not an assessable unit of toxicological information if that fact remains unqualified by other experimental data.

References

1.   Ashby, J. Mutagenesis 1986, 1, 3-16.
2.   Ashby, J.; Tennant, R. W. Mutation Res. 1988, 204, 17-115.
3.   Clayson, B. D. Mutation Res. 1987, 185, 243-271.

RECEIVED September 1, 1989

Chapter 13

Epidemiological Studies of Cancer and Pesticide Exposure

Allan H. Smith and Michael N. Bates

Department of Biomedical and Environmental Health Sciences, School of Public Health, University of California, Berkeley, CA 94720

Epidemiological studies concerning human cancer
risk and pesticides were reviewed. The most
persuasive evidence for a human cancer risk
was for the inorganic arsenic pesticides. There
was reasonably consistent evidence of an increase
in lung cancer risk for professional pesticide
applicators, although it was not clear that
this could be attributed to pesticides. Several
studies have found farmers to be at increased risk
for multiple myeloma, non-Hodgkins lymphoma and
leukemia, but no clear association with exposure
to pesticides has emerged. Some studies,
particularly in Sweden, have reported strong
associations between exposure to phenoxy herbicides
and malignant lymphoma and soft tissue sarcoma.
However, these associations have generally not been
supported by studies undertaken elsewhere and their
biological plausibility is questionable. Few
studies have examined the carcinogenic potential of
DDT and other organochlorine insecticides. There is
a need to do more epidemiological studies, particularly of the most highly exposed group - pesticide
applicators.

Since the discovery in the Eighteenth Century of scrotal cancer
in young chimney sweeps caused by exposure to soot,
epidemiological studies have identified a considerable number of
human carcinogens. The list includes cigarette smoke, asbestos,
benzene, vinyl chloride, nickel, chromium, cadmium, alcohol,
ionizing radiation, radon, benzidine and arsenic ([1]). Many of
these discoveries have come from studies of workers in various
industries.

Epidemiology is an observational science and, therefore,
causal inference involves a somewhat different approach to that
which is generally used in laboratory-based sciences. Since the
epidemiologist cannot usually conduct controlled experiments,

0097–6156/89/0414–0207$06.00/0

© 1989 American Chemical Society

bias in the ascertainment of the relationship between exposure
and cancer may arise due to errors in exposure assessment or in
ascertainment of cases or from confounding by other causal
factors. For these reasons, epidemiologists use several criteria
for causal inference in assessing exposure-disease relationships.
Foremost amongst these are the strength of an observed
association (the magnitude of the relative risk), the likelihood
that the association is spurious, the biological plausibility of
the association and its consistency from study to study. The
latter is a particularly important criterion, since it is
unlikely that a spurious association would be found in several
different studies, especially if they are conducted by different
investigators on different populations.

The purpose of this paper is to review epidemiological
studies concerning pesticides and human cancer. The review
commences with arsenical pesticides since they present the
strongest evidence for a group of pesticides causing human
cancer. No other pesticide or group of pesticides has been
convincingly identified as a cause of human cancer. The second
strongest association is that reported between pesticide
applicators and lung cancer, although no specific pesticides have
been identified as causal. Studies in different parts of the
world have reported excess rates of lymphopoietic cancers among
farmers and agricultural workers. However, our review of these
studies suggests that pesticide exposure is unlikely to be the
cause.

The phenoxy herbicides, including 2,4-D and 2,4,5-T (which
was invariably contaminated with 2,3,7,8-tetrachlorodibenzo-p-
dioxin formed during the manufacturing process), have been
postulated as possible causes of increased risks of malignant
lymphoma and soft tissue sarcoma among farmers. However,
considered collectively, the available studies provide only weak
evidence that these herbicides may be carcinogenic to humans.

ARSENICAL PESTICIDES

It has been well established by a variety of epidemiologic
studies that inorganic arsenic can cause human cancer. This is
in spite of the fact that animal studies have not yet shown
arsenic to be carcinogenic. Evidence comes from studies of
smelter workers, which have demonstrated increased lung cancer
risks (2,3). In addition, ingestion of arsenic has been shown
to cause cancer of the skin and other sites in studies of
exposure to naturally occurring arsenic in drinking water in
Taiwan (4,5), and studies of patients who have consumed medicinal
arsenic (6,7).

Studies involving arsenical pesticides include that of
Mabuchi et al (8) who studied mortality of a cohort of 1393
persons who had worked during the period 1946-77 in a plant in
Baltimore, Maryland, which manufactured and packaged various
pesticides, including inorganic arsenicals. Key results are
shown in Table I. There were 47 cancer deaths with 39.4 expected
based on Baltimore City rates (SMR=1.19; 90% confidence
interval: 0.92-1.52). Among the cancer deaths, 23 were from lung
cancer, 13.7 being expected (SMR=1.68; 90% CI: 1.15-2.38), with

the greatest risk found for workers with 25 or more years of employment (SMR=6.78; 90% CI: 3.54-11.83). These findings demonstrate a significant increase in lung cancer risk well beyond that which could be attributed to confounding by smoking.

Table I. Cancer Findings from Two Studies Involving Persons Exposed to Arsenical Pesticides

Reference		SMR	90% confidence interval
Mabuchi et al (8)			
manufacture	all deaths	0.97	0.86-1.09
	all cancer	1.19	0.92-1.52
	lung cancer	1.68	1.15-2.38
25+ years employed	lung cancer	6.78	3.54-11.83
Luchtrath (9)		OR	
arsenic exposed wine growers	lung cancer	14.7	9.0-24.0

Luchtrath (9) reported a study of mortality among German wine-growers affected by chronic arsenic poisoning, either through use of arsenic-containing insecticides or from drinking Haustrunk, a wine-substitute made from an aqueous infusion of already-pressed grapes and having a high arsenic content. He studied post-mortem findings for 163 winegrowers who had been diagnosed as suffering from chronic arsenic poisoning, comparing them with findings for another post mortem series of 163 men of similar age. Thirty of the arsenic-poisoned cases had skin cancer compared with none of the controls. Actual relative risk estimates were not given but various estimates can be calculated from data presented in the paper. Of the arsenic-poisoned growers 108 had lung cancer compared with only 14 of the controls. This gives an estimate of the relative risk for lung cancer of 20.90 (90% CI: 13.12-33.3). This estimate is extraordinarily high and, in view of the limited details on the process of selecting both cases and controls, is open to the suspicion of selection bias, particularly with the cases.

Such a bias may be less likely to be present in the mortality data obtained from the trade association, which was also presented by Luchtrath. Among 417 deaths of wine growers, 242 had lung carcinomas. Using the same post-mortem comparison group as above leads to a relative risk estimate for lung cancer of 14.7 (Table I). This estimate is lower than that calculated before, but still gives clear evidence of markedly increased lung cancer risks, well beyond anything which might be attributable to smoking, or other sources of bias.

Although epidemiological studies of people exposed to arsenical pesticides are limited in number, the two reviewed above provide evidence of increases in cancer rates attributable

to arsenic. The evidence is convincing because of the strength
of the relationships (high relative risk estimates) and because
other studies have convincingly demonstrated that inorganic
arsenic is a human carcinogen. Although there is a lack of
epidemiological studies to justify a definitive statement on the
matter, organic arsenical pesticides are not necessarily
carcinogenic. If they are, it would be reasonable to expect that
they would be be less potent than inorganic arsenic pesticides
since organic arsenicals are generally excreted much more rapidly
and tissue concentrations would, therefore, be lower than for
inorganic arsenic exposure.

PESTICIDE APPLICATORS AND LUNG CANCER

Considering the widespread concern about possible cancer risks
from pesticides, there have been few studies of professional
pesticide applicators, the most highly exposed group. The most
consistent findings from the studies which have been published
concern increased lung cancer risks (Table II). However, the
relative risk estimates were not sufficiently high to totally
discount smoking differences as an explanation.

Table II. Lung Cancer Standardized Mortality Ratios (SMRs)
           from Pesticide Applicator Mortality Studies

Reference		SMR	90% confidence interval
Wang & MacMahon (10)	all deaths	0.84	0.76-0.92
	all cancer	0.83	0.66-1.04
	lung cancer	1.15	0.79-1.61
MacMahon et al (11)	all deaths	0.98	0.93-1.03
	all cancer	1.11	1.00-1.23
	lung cancer	1.35	1.14-1.58
Barthel (12)	all cancer	0.82	0.55-1.20
	lung cancer	1.08	0.62-1.75
Riihimaki et al (13)	all deaths	0.74	0.61-0.89
	all cancer	0.82	0.55-1.20
	lung cancer	1.08	0.62-1.75
Blair et al (14)	all deaths	1.03	0.94-1.12
	all cancer	1.14	0.94-1.37
	lung cancer	1.35	1.00-1.80
	lung cancer	2.89	1.44-5.21

The first study was by Wang and MacMahon (10) and found a
small increase in lung cancer mortality rates among a cohort of
over 16,000 professional pesticide applicators. The updated
report of the same cohort (11) confirmed an increase in lung
cancer rates (SMR=1.35; 90% CI: 1.14-1.58). The excess risk

was confined to pesticide applicators who had not been employed
in termite control work (SMR=1.58; 90% CI: 1.29-1.90). The
animal carcinogens chlordane and heptachlor, widely employed for
termite control, were therefore thought unlikely to be involved.
It was also noted that the excess lung cancer cases were largely
confined to applicators who had been employed less than 5 years
in such work. This is evidence against a causal role for
pesticides.

A mortality study by Barthel of 1658 German men who had been
employed in agricultural work involving use of pesticides found
50 lung cancer cases when 27.5 were expected (SMR=2.1; 90%
CI: 1.6-2.8) (Table II) (12). A dose-response relationship was
demonstrated by years of exposure. However it was not possible
to link the increased lung cancer risk to any particular
pesticide or group of pesticides. A random sample of study
subjects did not smoke more than a general population sample.

A study of Finnish pesticide applicators by Riihimaki (13)
(Table II) focused on the phenoxy herbicides 2,4-D and 2,4,5-T.
There was no evidence of increased cancer risks from this
relatively small study. Twelve cases of lung cancer were found,
with 11.1 expected (SMR=1.08; 90% CI: 0.62-1.75).

The final study in this section involved 3827 male licensed
structural pesticide applicators in Florida (14). The lung
cancer SMR was 1.35, increasing to 2.89 for those who had been
licensed for at least 20 years (Table II). A relative risk of
this magnitude is unlikely to be due solely to confounding by
smoking. However, there was some increase in deaths from
emphysema, particularly in applicators who had been licensed for
10 years or more, indicating that smoking cannot be entirely
discounted as a source of bias. Again, it was not possible to
identify specific pesticides which might have been involved.

Taken as a group, the studies of pesticide applicators
provide limited, but reasonably consistent, evidence of increased
lung cancer risks, particularly if the Finnish study is regarded
as a study of a special type of application (agricultural use of
phenoxy herbicides). However, it is not certain that pesticides
are implicated and, if they are, the studies give no indication
as to which pesticides might be involved. Nonetheless, these
findings highlight the need for studies of applicators with at
least some individual exposure data, however crude. It is
possible that there might be considerable increases in lung
cancer risks attributable to particular pesticides, but in a
manner that leaves the overall relative risks rather low.

INCREASED CANCER RATES IN FARMERS

Increased rates of hemolymphopoietic cancers among farmers have
been reported in studies from several different countries (15-
29). Table III gives overall results for studies concerning
multiple myeloma, non-Hodgkin's lymphoma, and leukemia. If the
reported excesses were attributable to pesticides, then one would
expect to find considerable increased risks among those farmers
and agricultural workers most heavily exposed. Yet no clear
associations have emerged. It, therefore, seems unlikely that

Table III. Relative Risk Estimates for Farming and Multiple
            Myeloma (MM), non-Hodkin's Lymphoma (NHL) and
            Leukemia (L) with 95% Confidence Interval

Reference	Location		RR estimate	Confidence Interval
Fasal (19)	California	MM	1.00	0.74-1.32
		NHL	0.79	0.53-1.14
		L	1.14	0.94-1.36
Milham (25)	Oregon & Wash.	MM	1.79	1.13-3.00
	Oregon	L	1.35	1.00-1.78
	Washington	L	1.36	1.13-1.70
Blair (28)	Nebraska	L	1.25	1.05-1.49
Buesching (18)	Illinois	NHL	2.65	< 0.05
		L	2.00	< 0.05
Cantor (26)	Wisconsin	NHL	1.22	0.98-1.51
Burmeister (20)	Iowa	L	1.24	1.09-1.42
Burmeister (15)	Iowa	MM	1.48	< 0.05
		NHL	1.26	< 0.05
Gallagher (16)	Canada	MM	2.2	1.2-4.0
Cantor (27)	Wisconsin	MM	1.4	1.0-1.8
Delzell (23)	N. Carolina	MM	0.9 (W)	0.7-1.2
			1.0 (B)	0.9-1.1
		NHL	1.0 (W)	0.6-1.5
			1.2 (B)	0.4-3.9
		L	0.8 (W)	0.7-1.1
			1.9 (B)	1.1-3.1
Schumacher (24)	Utah	NHL	1.3	0.9-2.3
Pearce (22)	New Zealand	MM	2.22	1.31-3.75
		NHL	1.38	0.94-2.03
Steineck (21)	Sweden	MM	1.2	1.09-1.33
Wiklund (29)	Sweden	NHL	0.97	0.89-1.06

Note: The Delzell study gave separate results for whites (W)
       and blacks (B).

the fairly consistent, but small, excess of these cancers among farmers is primarily attributable to pesticides.

The most extensively investigated causal hypothesis involves the phenoxy herbicides. However, the studies which address the phenoxy herbicide hypothesis, reviewed in the next section, do not support the view that exposure to these chemicals is the explanation of the findings among farmers.

One possible explanation of the elevated cancer risks in Table III is infection by zoonotic oncoviruses. However, this hypothesis is not yet supported by evidence of human seropositivity to such viruses, although a number of New Zealand studies have provided evidence of an elevated risk for non-Hodgkins lymphoma and soft tissue sarcoma in abattoir workers (30-33). The significance of this finding is that such workers are not likely to have been greatly exposed to pesticides, whereas they will have had ample opportunity for exposure to animal viruses.

MALIGNANT LYMPHOMA AND PHENOXY HERBICIDES

The first study to report an association between malignant lymphoma and phenoxy herbicides was conducted in Sweden (34) (Table IV). A relative risk estimate of 4.8 was obtained for persons who had ever sprayed phenoxy herbicides, mainly 2,4,5-T and 2,4-D. Relative risk estimates were 4.3 for those with less than 90 days total exposure and 7.0 for those with more than 90 days exposure. These were remarkable findings since, with one exception, no occupational exposure had previously been reported to cause such increases in cancer risk after such short exposures. The particular exception was a study of asbestos-factory workers in which relative risks of almost 2 were reported for 3 months of very heavy exposures (35).

A relative risk of 4.3 from very short-term pesticide exposures would be expected to be associated with an overall increase in cancer risk amongst agricultural workers. However, Wiklund et al (29) found a relative risk of 0.97 for non-Hodgkins Lymphoma (NHL) and land/animal husbandry (95% CI: 0.89-1.06) for non-Hodgkins Lymphoma (NHL) in a study of 354,620 Swedish male agricultural and forestry workers involving linkage between census and national tumor registry data (Table IV). Also, no increased risks were found in forestry workers. Moreover, in a study of 20,245 registered Swedish pesticide applicators (36) the relative risk for NHL was found to be 1.16 (95% CI: 0.66-1.86) in the group which was first licensed in 1965/66. No satisfactory explanation has been put forward to explain the discrepancy between the results of the case-control study and the two cohort studies, and one is left with questions concerning possible biases in the case-control study which reported the initial association.

In contrast to Sweden, a study in New Zealand reported an overall increase in rates of malignant lymphoma in farmers (Table III) (22). However, a case-control study revealed no association with phenoxy herbicide use, even though 2,4,5-T and 2,4-D have been used extensively in New Zealand since the late 1940s (31). The relative risk estimates for ever using phenoxy herbicides was

1.3 when using other cancer controls, and 1.0 when using general population controls (Table IV).

Table IV. Studies Concerning Non-Hodgkins's Lymphoma and
Exposure in the Use or Manufacture of
Phenoxy Herbicides

Reference	RR	90 or 95% confidence interval
SWEDEN		
Hardell et al (34)	4.8	2.9-8.1
Wiklund et al (36)	1.01	0.6-1.5
NEW ZEALAND		
Pearce et al (22)	1.3	0.7-2.5
	1.0	0.5-2.0
UNITED STATES		
Hoar et al (37)	2.3	1.3-4.3
Woods et al (38)	1.1	0.8-1.4
>15 years	1.7	1.0-2.8
Hoar et al (39)	1.5	0.9-2.4
>20 days/year	3.3	0.5-22.1
PHENOXY HERBICIDE MANUFACTURING		
Lynge (40)	1.3	0.5-2.7
Ott et al (41)	1.9	0.6-4.5

A study in Kansas (37) has produced evidence supporting an association between exposure to 2,4-D and malignant lymphoma. A relative risk of 2.3 (95% CI: 1.3-4.3) was found for those who had used phenoxy herbicides (almost all 2,4-D) (Table IV). However, a relative risk of 6.0 (95% CI: 1.9-19.5) was found for those who had used any herbicide (not necessarily 2,4-D) for more than 20 days per year. This association was largely confined to early years of herbicide use when no independent confirmation of exposure was available. There was no association with number of years of herbicide use after adjustment for days of herbicide use per year, which tends to reduce the likelihood of a real association.

A subsequent study in Washington State involving 576 cases of NHL reported a relative risk of 1.07 (95% CI: 0.8-1.4) associated with any past exposure to phenoxy herbicides (38). This estimate increased to 1.7 (95% CI: 1.04-2.8) for exposure of more than 15 years during the period prior to 15 years before diagnosis (Table IV).

In a case control study of NHL in Nebraska, Hoar Zahm et al (39) examined the association with exposure to various pesticides. A small increase in NHL (OR=1.5; 95% CI: 0.9-2.4) was associated with use of 2,4-D. The odds ratio increased to 3.3 (95% CI: 0.5-22.1) for exposure to 2,4-D of more than 20 days per year. Significant associations with use of chlordane, diazinon, dyfonate and malathion were also found.

Two studies have reported malignant lymphoma mortality for the chemical manufacturing industry (Table IV). Lynge (40), in a study of cancer incidence in a cohort of 4,459 persons who had been involved in phenoxy herbicide manufacture in Denmark, found 7 cases of malignant lymphoma among males, whereas 5.4 would have been expected (RR=1.30; 95% CI: 0.52-2.69). None of the 7 cases were recorded as having participated in the manufacture or packaging of phenoxy herbicides.

Ott et al (41) in a study of a cohort of 2192 workers who had manufactured chlorophenols (and, therefore, had potential exposure to chlorinated dioxins) found 5 deaths due to malignant lymphoma, when 2.6 would have been expected (SMR=1.92; 95% CI: 0.62-4.49). However, there was no increasing trend for risk associated with increasing opportunity for exposure.

Taken overall, studies to date have not confirmed an association between phenoxy herbicides and malignant lymphoma. In fact, it seems unlikely that phenoxy herbicides are the explanation for increased lymphoma rates among farmers reported in several studies. For example, phenoxy herbicides are not associated with NHL in New Zealand, where there is an excess NHL mortality among farmers, while the major evidence for an association with phenoxy herbicides comes from Sweden where there is no overall excess of NHL mortality in agricultural workers. As discussed below, the association also lacks biological plausibility.

SOFT TISSUE SARCOMA AND PHENOXY HERBICIDES

The hypothesis that phenoxy herbicides are a cause of soft tissue sarcoma (STS) also commenced in Sweden. Three case-control studies conducted by the same investigators reported relative risks of 3.3 to 5.7 for such an association (Table V) (42-44). However, as with malignant lymphoma, it is difficult to explain such dramatic risks associated with short term exposures. Moreover, Wiklund and Holm (45) found no overall risk increase from STS among Swedish agriculture and forestry workers (RR=0.9; 95% CI: 0.8-1.0).

In the study by Lynge, (40) mentioned above, of workers involved in the manufacture of phenoxy herbicides in Denmark, five male STS cases were found compared to 1.84 expected (RR=2.72; 95% CI: 0.88-6.34). However, only one of the patients had worked in the manufacture and packaging of phenoxy herbicides, one probably had a hereditary predisposition to neurofibrosarcoma and another appeared to have been diagnosed with STS before beginning employment at the plant. Moreover, three of the patients had been employed for 3 months or less.

Studies in New Zealand, where phenoxy herbicides, particularly 2,4,5-T, have been extensively used, have not found evidence for a relationship between use of these herbicides and STS (Table V) (46-47).

Studies in the U.S. have also not found an association with STS (37-38), although one study did find a relative risk of 2.8 associated with having a Scandinavian surname (95% CI: 0.5-15.6) (38). In addition, there has been no consistent link with STS in Vietnam Veterans, some of whom were exposed to the phenoxy

Table V.   Results of Studies Concerning Soft Tissue Sarcoma
and Exposure to Phenoxy Herbicides

Reference	RR	90 or 95% confidence interval
SWEDEN		
Hardell and Sandstrom (42)	5.7	2.9-11.3
Eriksson et al (44)	6.8	2.6-17.3
Wiklund and Holm (45)	0.9	0.8-1.0
Hardell and Eriksson (43)	3.3	1.4-8.1
NEW ZEALAND		
Smith et al (46)	1.6	0.7-3.3
Smith and Pearce (47)	0.7	0.3-1.5
Combined studies	1.1	0.7-1.8
UNITED STATES		
Hoar et al (37)	0.9	0.5-1.6
Woods et al (38)	0.8	0.5-1.2
Swedish surnames	2.8	0.5-15.6
DENMARK		
Lynge (40)	2.72	0.88-6.34
AGENT ORANGE, U.S. VIETNAM VETERANS		
Greenwald et al (48)	0.53	0.21-1.31
Lawrence et al (49)	1.15	0.61-2.17
Kang et al (51)	0.83	0.63-1.09
Kang et al (50)	0.85	0.54-1.36
Military Region 3	8.64	0.77-111.84
Lathrop et al (54)	1/1016, Ranch Hand	
	1/1293, non-Ranch Hand	
Kogan and Clapp (55)	5.16	2.4-11.1

herbicide mixture Agent Orange (Table V) (48-52). Elevated rates have been reported for Massachusetts Veterans (53) and in a study involving West Virginia Veterans (which involved only 3 exposed cases) (52). However, elevated risks were not found in the larger studies (48, 49, 51, 54). In addition, the servicemen who served in Operation Ranch Hand had much higher exposures to Agent Orange than most Veterans, but so far no excess of STS has been found in this group (Table V) (54).

One case-control study in Italy has reported an increased risk of STS for women who worked in rice fields (55). While it has been suggested that this study provides evidence implicating phenoxy herbicides, at best the evidence is very weak, since the association was based on 3 women who worked in the rice fields during the 1950s. There is no suggestion that they had any direct contact with phenoxy herbicides, although it was postulated that they may have had indirect exposure in the rice paddies from previously-applied herbicides.

When the STS studies are considered together, the evidence implicating phenoxy herbicide exposure is weak. The studies which provide the strongest evidence are from Sweden, and the absence of an overall risk increase in agriculture and forestry in that country has not been satisfactorily explained.

BIOLOGICAL PLAUSIBILITY OF PHENOXY HERBICIDES CAUSING CANCER

The initial hypothesis that phenoxy herbicides were a cause of human cancer was prompted by the fact that 2,4,5-T was contaminated with 2,3,7,8-TCDD, a potent animal carcinogen. However, the epidemiologic evidence does not support this hypothesis. Before about 1970 2,4,5-T was contaminated with about 1 ppm of 2,3,7,8-TCDD (47)(56). Studies indicate that a 2,4,5-T sprayer generally absorbs less than 0.2 mg/kg/workday of 2,4,5-T (57). Assuming a directly proportionate amount of 2,3,7,8-TCDD contaminant is concurrently absorbed, then 0.2 ng/kg/day of TCDD would be absorbed at a 2,4,5-T contamination level of 1 ppm. Making the further assumption that the human cancer potency of TCDD is similar in man to animals, it can be shown that it is unlikely that increases in human cancer risk would be epidemiologically detected (58).

It is also noteworthy that 2,3,7,8-TCDD has a very long half-life in humans, and elevated levels have been reported many years after exposure in the blood of Ranch Hand servicemen and workers who had been involved in trichlorophenol manufacture (59). However, studies in Sweden involving cases of soft tissue sarcoma and malignant lymphoma have not found increased levels of 2,3,7,8-TCDD in fat (60). This provides evidence against the implication of 2,4,5-T, since significant exposure to this herbicide should lead to long term increases in levels of 2,3,7,8-TCDD in adipose tissue.

It is possible that the phenoxy herbicides themselves may cause cancer. However, this hypothesis lacks biological plausibility since the uncontaminated phenoxy herbicides are not animal carcinogens, nor are they genotoxic. The example of arsenic shows that one should not automatically conclude that non-animal carcinogens would not cause cancer in humans.

However, the human carcinogenicity findings for arsenic are convincing, in part because exposures were high and over a long period, producing other toxic effects as well as cancer. By contrast, those studies which have produced increased cancer risks with phenoxy herbicide exposure have involved short low intensity exposures. This tends to reduce the biological plausibility of such findings.

ORGANOCHLORINE PESTICIDES

It is well established that organochlorine pesticides, such as DDT, dieldrin, aldrin, heptachlor and chlordane, can cause tumors in rodents, particularly liver cancer in mice. A number of investigations have sought to determine whether these pesticides are carcinogenic to humans.

Considering its relatively long history of widespread use, it is remarkable that epidemiological studies of cancer risk focused on DDT exposure are almost non-existent. Ditraglia et al (61) studied mortality in workers from four organochlorine pesticide manufacturing plants, one of which had solely manufactured DDT since 1947. Cohorts were defined as all workers who had been employed for at least six months prior to 1965 and vital status was ascertained up to 1976. For the DDT plant a total of 6 cancer deaths were found leading to the calculation of an SMR of 0.68 (95% CI: 0.25-2.47) for all cancers combined. The utility of this study is limited because of the small number of workers involved.

Ortelee (62) studied a total of 40 men employed in the manufacture or formulation of DDT for up to 8 years and concluded that, with the exception of some minor skin and eye irritations, no illness was associated with exposure to DDT in this group of workers.

Laws et al (63) studied a group of 35 men who had been employed in the production of DDT for more than five years. The authors concluded that they had found no ill effects, including cancer, attributable to exposure to DDT.

Neither of the latter two studies has either sufficient subjects or sufficient time of follow-up for any conclusions to be drawn regarding the human cancer risk of DDT.

In a retrospective cohort study of 1403 male pesticide workers employed in the manufacture of chlordane and heptachlor between 1946 and 1976, Wang and MacMahon (64) found no statistically significant excess of cancer mortality either overall (SMR=82; 95% CI: 54-120) or at any particular cancer site. However, the SMR for lung cancer was 1.34 (95% CI: 73-228), but there was no pattern in relation to either exposure level or latency which would suggest a causal role for heptachlor or chlordane.

Ditraglia et al (61) also studied mortality experience in workers from the two plants involved in the study by Wang and MacMahon as well as two additional plants, one of which had manufactured aldrin, dieldrin and endrin. The other, discussed above, had manufactured only DDT. Mortality from all malignant neoplasms was less than expected, with SMR's from 0.73 to 0.91 for the three plants which had not manufactured DDT. Although

SMR's for some individual cancer sites exceeded 1, the observed numbers of cancers were small.

Austin et al (65) performed a case-control study using 80 hepatocellular cancer patients and 146 matched hospital controls. Relative risks for exposures to herbicides and other pesticides were 1.0 and 2.1 (95% CI: 0.6-6.9), respectively.

In conclusion, epidemiological evidence has not indicated an association of organochlorine pesticides with increases in human cancer incidence rates. However, in regard to DDT at least, there is no adequate epidemiological evidence on which to base a judgement of human carcinogenicity. Any conclusion regarding the other organochlorine pesticides must be tempered by the low numbers of exposed subjects (and consequent low power to detect moderate increases in relative risk) in the studies which have been performed so far.

[Note added in proof: A recently published study reported results of a prospective follow-up study of 1,708 adult residents of Charleston, S. Carolina (69). Study subjects were not selected on the basis of any particular occupational exposure. Blood samples were taken in 1974-75 and analysed for DDT and DDE. Vital status of the cohort was ascertained in 1984-85. No elevation of risk for overall cancer mortality was found, although there was a non-significant increasing trend for respiratory cancer deaths with increasing blood DDT level].

AMITROLE

Amitrole, a triazole herbicide and plant growth regulator, has been shown in toxicological studies to be carcinogenic to both rats and mice, but not hamsters (66). Axelson and Sundell (67) found an SMR for cancer of 3.6 (4 observed, 1.11 expected; 90% CI: 1.2-8.2) for exposure to amitrole with a latent period of at least 5 years in a cohort of Swedish railway workers.

In a further study which extended the follow-up period by 6 years Axelson et al (68) obtained an SMR of 1.5 for amitrole exposure (5 observed, 3.34 expected; 90% CI: 0.6-3.1). The observed tumors were of a range of types, with two being lung cancers. The authors also noted excess cancers for those exposed to both phenoxy herbicides and amitrole and suggested that the phenoxy herbicides may be responsible for the observed cancers.

CONCLUSION

With the exception of the inorganic arsenicals, epidemiological studies have not consistently demonstrated increased cancer risks with the use of pesticides, although findings of increased lung cancer risks amongst applicators warrant further study and further investigatation of phenoxy herbicides is needed. The epidemiological studies to date are reassuring for the general population, which is exposed to much lower levels of pesticides than the occupational groups studied. However, studies of pesticide users are very limited and priority should be given to studies among professional pesticide applicators, the most highly exposed group, including prospective cancer incidence surveillance.

LITERATURE CITED

1. Doll, R.; Peto, R. J. Natl. Cancer Inst. 1981, 66,
    1192-308.
2. Axelson, O.; Dahlgren, E.; Jansson, C. D.; Rehnlund, S. O.
    Br. J. Ind. Med. 1978, 35, 8-15.
3. Enterline, P. E.; March, G. M. Am. J. Epidemiol. 1982, 116,
    895-911.
4. Tseng, W-P. Environ. Health Persp. 1977, 19, 109-19.
5. Tseng, W. P.; Chu, H. M.; How, S. W.; Fong, J. M.; Lin, C.
    S.; Yeh, S. JNCI 1968, 40, 453-63.
6. Cusick, J.; Evans, S.; Gillman, M.; Price, Evans D. A. Br.
    J. Cancer 1982, 45, 904-11.
7. Sommers,S. C.; McManus, R. G. Cancer 1953, 6, 347-59.
8. Mabuchi, K.; Lilienfeld, A. M.; Snell, L. M. Prev. Med.
    1980, 9, 51-77.
9. Luchtrath H. J. Cancer Res. Clin. Oncol. 1983, 105, 173-82.
10. Wang, H. H.; MacMahon, B. J. Occup. Med. 1979, 21, 741-44.
11. MacMahon, B.; Monson, R. R.; Wang, H. H.; Zhang, T. JOM
    1988, 30, 429-32.
12. Barthel, E. J. Toxicol. Environ. Health 1981, 8, 1027-40.
13. Riihimaki, V.; Asp, S.; Pukkala, E.; Hernberg, S.
    Chemosphere 1983, 12, 779-84.
14. Blair, A.; Grauman, D. J.; Lubin, J. H.; Fraumeni, J. F.
    JNCI 1983, 71, 31-7.
15. Burmeister, L. F.; Everett, G. D.; Lier, S. F.; Isaacson, P.
    Am. J. Epidemiol. 1983, 118, 72-7.
16. Gallagher, R. P.; Spinelli, J. J.; Elwood, J. M.; Skippen,
    D. H. Br. J. Cancer 1983, 48, 853-7.
17. Priester, W.A.; Mason, T. J. JNCI 1974, 53, 45-9.
18. Buesching, D. P.; Wollstadt, L. JNCI 1984, 72, 503.
19. Fasal, E.; Jackson, E. W.; Klauber, M. R. Am. J. Epidemiol.
    1968, 87, 267-74.
20. Burmeister, L. F.; Lier, S. F.; Isaccson, P. Am. J.
    Epidemiol. 1982, 115, 720-8.
21. Steineck, G.; Wiklund, K. Inter. J. Epi. 1986, 15, 321-5.
22. Pearce, N. E.; Smith, A. H.; Fisher, D. O. Am. J.
    Epidemiol. 1985, 121, 225-37.
23. Delzell, E.; Grufferman, S. Am. J. Epidemiol. 1985, 121,
    391-402.
24. Schumacher, M. C. J. Occup. Med. 1985, 27, 580-4.
25. Milham, S. Am. J. Epidemiol. 1971, 94. 307-10.
26. Cantor, K. P. Int. J. Cancer 1982, 29, 239-47.
27. Cantor, K. P.; Blair, A. JNCI 1984, 72, 251-55.
28. Blair, A.; Thomas, T. L. Am. J. Epidemiol. 1979, 110, 264-
    73.
29. Wiklund, K.; Lindefors, B-M. ; Holm, L-E. Br. J. Ind.
    Med. 1988, 45, 19-24.
30. Smith, A. H.; Pearce, N. E.; Fisher, D. O.; Giles, H. J.;
    Teague, C. A; Howard, J. K. JNCI 1984, 73, 1111-17.
31. Pearce, N. E.; Smith, A. H.; Howard, J. K.; Sheppard, R. A.;
    Giles, H. J.; Teague, C. A. Br. J. Ind. Med. 1986, 43, 75-
    83.

32. Pearce, N. E.; Sheppard, R. A.; Howard, J. K.; Fraser, J.; Lilley, B. M. Am. J. Epidemiol. 1986, 124, 402-9.
33. Pearce, N. E.; Sheppard, R. A.; Smith, A. H.; Teague, C. A. Int. J. Cancer 1987, 39, 155-61.
34. Hardell, L.; Eriksson, M.; Lenner, P.; Lundgren, E. Br. J. Cancer 1981, 43, 169-76.
35. Seidman, H.; Selikoff, I. J.; Hammond, E. C.; Ann. NY. Acad. Sci. 1979, 330, 61-90.
36. Wiklund, K.; Dich, J.; Holm, L-E. Br. J. Cancer 1987, 56, 505-8.
37. Hoar, S. K.; Blair, A.; Holmes, F. F.; Boysen, C. D.; Robel, R. J.; Hoover, R.; Fraumeni, J. F. JAMA 1986, 256, 1141-7.
38. Woods, J. S.; Polissar, L.; Severson, R. K.; Heuser, L. S.; Kulander, B. G. JNCI 1987, 78, 899-910.
39. Hoar, Zahm S.; Wiesenberger, D. D.; Babbitt, P. A.; Saal, R. C.; Cantor, K. P.; Blair, A. Presented at the Society for Epidemiological Research, 1988.
40. Lynge, E. Br. J. Cancer 1985, 52, 259-70.
41. Ott, M. G.; Olson, R. A.; Cook, R. R.; Bond, G. G. J. Occup. Med. 1987, 29, 422-9.
42. Hardell, L.; Sandstrom, A. Br. J. Cancer 1979, 39, 711-17.
43. Hardell, L.; Eriksson, M. Cancer 1988, 62.
44. Eriksson, M; Hardell, L.; Berg, N. O.; Moller, T.; Axelson, O. Br. J. Ind. Med. 1981, 38, 27-33.
45. Wiklund, K.; Holm, L. E. JNCI 1986, 76, 229-34.
46. Smith, A. H.; Fisher, D. O.; Pearce, H.; Teague, C. A. Community Health Stud 1982, 6, 114-19.
47. Smith AH, Pearce NE. Chemosphere 1986, 15 1795-8.
48. Greenwald, P.; Kovasznay, B.; Collins, D. N.; Therriault, G. JNCI 1984, 73, 1107-9.
49. Lawrence, C. E.; Reilly, A. A.; Quickenton, P.; Greenwald, P.; Page, W. F.; Kuntz, A. J. AJPH 1985, 75, 277-9.
50. Kang, H.; Enziger, F.; Breslin, P.; Feil, M.; Lee, Y.; Shepard, B. JNCI 1987, 79, 693-9.
51. Kang, H. K.; Weatherbee, L.; Breslin, P. P.; Lee, Y.; Shepard, B. M. J. Occup. Med. 1986, 28, 1215-8.
52. West Virginia Health Department. In West Virginia Vietnam-era Veterans Mortality Study, 1986.
53. Kogan, M. D.; Clapp, R. W. Int. J. Epidemiol. 1988, 17, 39.
54. Lathrop, G. D.; Machado, S. G.; Karrison, T. G.; Grubbs, W. D.; Thomas, W. F.; Wolfe, W. H.; Michalek, J. E.; Miner, J. C.; Peterson, M. R.; Ogershok, R. W. Science Applications International Corporation, 1987.
55. Vineis, P.; Terracini, B.; Ciccone, G.; Cignetti, A.; Colombo, E.; Donna, A.; Maffi, L.; Pisa, R.; Ricci, P.; Zanini, E.; Comba, P. Scand. J. Work Environ. Health 1986, 13, 9-17.
56. Norstrom, A.; Rappe, C.; Lindahl, R; Buser, H. R. Scand. J. Work Environ. Health 1979, 5, 375-378.
57. Leng, M. L.; Lavy, T. L.; Ramsey, J. C.; Braun, W. H. American Chemical Society 1982, p 133-156.
58. Human Health Aspects of Environmental Exposure to Polychlorinated Dibenzo-p-dioxins and Polychlorinated Dibenzofurans, Universities Associated for Research and Education in Pathology, Inc., Bethesda, MD, 1988.

59. Centers for Disease Control. JAMA 1988, 259, 3533-5.
60. Nygren, M. Academic Thesis, University of Umea, Sweden, 1988.
61. Ditraglia, D.; Brown, D. P.; Namekata, T.; Iverson, N. Scand. J. Work Environ. Health 1981, 7, Suppl. 4, 140-46.
62. Ortelee, M. F. AMA Arch. Ind. Health 1958, 18, 433-40.
63. Laws, E. R.; Curley, A.; Biros, F. J.; Arch. Environ. Health 1967, 15, 766-75.
64. Wang, H, MacMahon, B. J. Occup. Med. 1979, 21, 745-48.
65. Austin, H.; Delzell, E.; Grufferman, S.; Levine, R.; Morrison, A. S.; Stolley, P. D.; Cole, P. JOM 1987, 29, 665-9.
66. International Agency for Research on Cancer, Vol. 41, France, 1986.
67. Axelson, O.; Sundell, L. Work Environ. Health 1974, 11, 21-8.
68. Axelson, O.; Sundell, L.; Andersson, K.; Edling, C.; Hogstedt, C.; Kling, H. Scand. J. Work Environ. Health 1980, 6, 73-9.
69. Austin, H.; Keil, J. E.; Cole, P. Amer. J. Pub. Health 1989, 79, 43-46.

RECEIVED June 28, 1989

Chapter 14

Pesticide Residues and Cancer Causation

Bruce N. Ames

Department of Biochemistry, University of California, Berkeley, CA 94720

There are many sources of natural mutagens and
carcinogens in the environment, such as natural toxic
chemicals present in all plants as defenses against
insects, chemicals formed on cooking or preparing food,
and mold products and other dietary components. Since
natural carcinogens appear to be extremely common,
priority setting is required to separate important from
trivial hazards. We discuss mechanisms of
carcinogenesis and the implications of this for doing
risk assessment. We also discuss reasons why animal
cancer tests cannot be used to predict absolute human
risks. Such tests, however, may be used to indicate
that some chemicals might be of greater concern than
others. Possible hazards to humans from a variety of
rodent carcinogens are ranked by an index that relates
the potency of each carcinogen in rodents to the
exposure in humans. This ranking suggests that
carcinogenic hazards from current levels of pesticide
residues (or water pollution) are likely to be of
minimal concern relative to the background levels of
natural substances, though one cannot say whether these
natural exposures are likely to be of major or minor
importance.

Many Chemicals are Carcinogens and Reproductive Toxins, and We Cannot
Eliminate All of Them

Over 50% of the chemicals tested to date in rats and mice have been
found to be carcinogens at the high doses administered (1,2), the
maximum tolerated dose (MTD). The exhaustive database of animal
cancer tests developed by my colleagues and me (3,4) listed 392
chemicals tested in both rats and mice at the MTD. Of these, 58% of
the synthetic chemicals and 45% of the natural chemicals were
carcinogens in at least one species (1,2). We concluded that the
proportion of chemicals found to be carcinogens is strikingly high, a
conclusion reached by others with smaller compilations. The earlier

0097–6156/89/0414–0223$06.00/0
© 1989 American Chemical Society

Innes _et al_. study (5) is sometimes cited to support the conclusion
that the proportion of carcinogens is low. The Innes study was a
much smaller dataset (120 chemicals, 11 positive), and the tests,
though appropriate for their time, used only one species and were
less thorough than modern tests (2).

Even when one considers that some chemicals are selected for
testing because they are suspicious, the high proportion of positives
is disturbing. From considerations of carcinogenesis mechanisms, it
is plausible that a high proportion of all chemicals we test in the
future, both natural and man made, will prove to be carcinogens (1;
see section on Extrapolating Risks).

High proportions of positives are also reported for teratogenic
tests. Fully one-third of the 2800 chemicals tested in laboratory
animals have been shown to induce birth defects at maximum tolerated
doses (6). Thus, it seems likely that a sizeable percentage of both
natural and man-made chemicals will be reproductive toxins when
tested at the MTD. The world is full of carcinogens and reproductive
toxins, and it always has been. The important issue is the human
exposure dose, and, fortunately, almost all of these are usually
tiny.

The major preventable risk factors for cancer causation, such as
tobacco, dietary imbalances (7-13), hormones (14), and viruses
(15,16), have been discussed by us (1,17-22) and others (7,14,23-25).

Man-Made Chemical Pollutants Do Not Appear to be Present in
Significant Amounts

We have attempted to address the issue of priority setting among
possible carcinogenic hazards (1). Since carcinogens differ
enormously in potency in rodent tests, a comparison of possible
hazards from various carcinogens ingested by humans must take this
into account. Our analysis makes use of an exhaustive database of
animal cancer tests (currently 3500 experiments on 975 chemicals)
(3,4) that calculates carcinogenic potency, the TD_{50}, essentially the
dose of the carcinogen to give half of the animals cancer. The TD_{50}
is close to the high dose (MTD) actually given and thus involves a
minimal extrapolation. To calculate our index of possible hazard we
express each human exposure (daily lifetime dose in mg/kg) as a
percentage of the rodent TD_{50} dose (mg/kg) for each carcinogen. We
call this percentage HERP (Human Exposure dose/Rodent Potency dose).
As rodent data are all calculated on the basis of lifetime exposure
at the indicated daily dose rate (3,4), the human exposure data are
similarly expressed as lifelong daily dose rates even though the
human exposure is likely to be less than daily for a lifetime. The
HERP values are not risk assessment, because it is impossible to
extrapolate to low doses (see section on Extrapolating Risks), but
are a way of comparing possible hazards of exposures to put them in
perspective and to set priorities (Table 1).

This analysis suggests that the amounts of pollutants that
humans are ingesting from pesticide residues or water pollutants
appear to be trivial relative to the background of natural, and
traditional (e.g., from cooking food) carcinogens (1,22).

Nature's Pesticides. Americans ingest in their diet at least 10,000
times more by weight of natural pesticides than of man-made pesticide

residues (22). These natural "toxic chemicals" have an enormous
variety of chemical structures, appear to be present in all plants,
and serve as protection against fungi, insects, and animal predators
(21,22). Though only a few dozen are found in each plant species,
they commonly make up 5 to 10% of the plant's dry weight (22). There
has been relatively little interest in the toxicology or
carcinogenicity of these compounds until quite recently, although
they are by far the main source of "toxic chemicals" ingested by
humans.

Most chemicals tested for carcinogenicity in rodent bioassays
are synthetic compounds; however, the proportion of positive tests is
about as high for natural pesticides as for synthetic chemicals
(roughly 30%). Since over 99.99% of the pesticides we ingest are
"nature's pesticides" (1,21,22), our diet is likely to be very high
in natural carcinogens. Their concentration is usually in parts per
thousand or more rather than parts per billion, as is usual for
synthetic pesticide residues or water pollutants (1). The known
natural carcinogens in mushrooms, parsley, basil, parsnips, fennel,
pepper, celery, figs, mustard, cabbage, broccoli, Brussels sprouts,
carrots, pineapple, and citrus juices are undoubtedly just the
beginning of the list of natural carcinogens, since so few of
"nature's pesticides" have been tested (1,21,22). For example, a
recent analysis (26) of lima beans showed an array of 23 natural
alkaloids (those tested have biocidal activity) that ranged in
concentration in stressed plants from 0.2 to 33 parts per thousand
fresh weight. None appear to have been tested for carcinogenicity or
teratogenicity.

Man-made Pesticide Residues. Intake of man-made pesticide residues
from food in the United States, including residues of industrial
chemicals such as polychlorinated biphenyls (PCBs), has been
estimated by FDA. They assayed food for residues of the 70 compounds
thought to be of greatest importance (27). The human intake averages
about 150 μg/day. Most (105 μg) of this intake is composed of three
chemicals (ethylhexyl diphenyl phosphate, malathion, and
chlorpropham) shown to be noncarcinogenic in tests in rodents (1).
Thus, the intake of carcinogens from residues (45 μg/day if all the
other residues are carcinogenic, which is unlikely) is extremely tiny
relative to the background of natural substances (1,22).

The latest figures from the FDA about actual exposures don't
include every known man-made pesticide, but they are a reasonable
attempt at doing so. In a recent NRC/NAS report, Regulating
Pesticides in Food (28), it is suggested that some of the pesticides
not covered by the FDA sampling, particularly those used on tomatoes,
should have their allowable limits lowered and presumably should be
added to the FDA sampling program. Nevertheless, the estimate of 45
μg of possibly carcinogenic pesticide residues consumed in a day is
likely to be a reasonable estimate, as is our conclusion that the
possible hazards from these residues are minimal compared to the
background of nature's pesticides. To put the amounts of man-made
pesticides in perspective (1), there are about 500 μg of carcinogens
in a cup of coffee (hydrogen peroxide and methylglyoxal), 185 μg of
carcinogenic formaldehyde in a slice of bread, about 2,000 μg of
formaldehyde in a cola, 760 μg of carcinogenic estragole in a basil
leaf, and a gram of burnt material from cooking our food.

Table 1. Ranking possible carcinogenic hazards

Possible hazard: HERP (%)	Daily human exposure	Carcinogen dose per 70-kg person	Potency of carcinogen: TD50 (mg/kg)	
			Rats	Mice
		Environmental pollution		
0.001*	Tap water, 1 liter	Chloroform, 83 µg (U.S. average)	(119)	90
0.004*	Well water, 1 liter contaminated (worst well in Silicon Valley)	Trichloroethylene, 2800 µg	(—)	941
0.0004*	Well water, 1 liter contaminated, Woburn	Trichloroethylene, 267 µg	(—)	941
0.0002*		Chloroform, 12 µg	(119)	90
0.0003*		Tetrachloroethylene, 21 µg	101	(126)
0.008*	Swimming pool, 1 hour (for child)	Chloroform, 250 µg (average pool)	(119)	90
0.6	Conventional home air (14 hour/day)	Formaldehyde, 598 µg	1.5	(44)
0.004		Benzene, 155 µg	(157)	53
2.1	Mobile home air (14 hour/day)	Formaldehyde, 2.2 mg	1.5	(44)
		Pesticide and other residues		
0.0002*	PCBs: daily dietary intake	PCBs, 0.2 µg (U.S. average)	1.7	(9.6)
0.0003*	DDE/DDT: daily dietary intake	DDE, 2.2 µg (U.S. average)	(—)	13
0.0004	EDB: daily dietary intake (from grains and grain products)	Ethylene dibromide, 0.42 µg (U.S. average)	1.5	(5.1)
		Natural pesticides and dietary toxins		
0.003	Bacon, cooked (100 g)	Dimethylnitrosamine, 0.3 µg	(0.2)	0.2
0.006		Diethylnitrosamine, 0.1 µg	0.02	(+)
0.003	Sake (250 ml)	Urethane, 43 µg	(41)	22
0.03	Comfrey herb tea, 1 cup	Symphytine, 38 µg (750 µg of pyrrolizidine alkaloids)	1.9	(?)
0.03	Peanut butter (32 g; one sandwich)	Aflatoxin, 64 ng (U.S. average, 2 ppb)	0.003	(+)
0.06	Dried squid, broiled in gas oven (54 g)	Dimethylnitrosamine, 7.9 µg	(0.2)	0.2
0.07	Brown mustard (5 g)	Allyl isothiocyanate, 4.6 mg	96	(—)
0.1	Basil (1 g of dried leaf)	Estragole, 3.8 mg	(?)	52
0.1	Mushroom, one raw (15 g) (Agaricus bisporus)	Mixture of hydrazines, and so forth	(?)	20,300
0.2	Natural root beer (12 ounces; 354 ml) (now banned)	Safrole, 6.6 mg	(436)	56
0.008	Beer, before 1979 (12 ounces; 354 ml)	Dimethylnitrosamine, 1 µg	(0.2)	0.2

2.8*	Beer (12 ounces; 354 ml)	Ethyl alcohol, 18 ml	9110	(?)
4.7*	Wine (250 ml)	Ethyl alcohol, 30 ml	9110	(?)
6.2	Comfrey-pepsin tablets (nine daily)	Comfrey root, 2700 mg	626	(?)
1.3	Comfrey-pepsin tablets (nine daily)	Symphytine, 1.8 mg	1.9	(?)
	Food additives			
0.0002	AF-2: daily dietary intake before banning	AF-2 (furylfuramide), 4.8 µg	29	(131)
0.06*	Diet Cola (12 ounces; 354 ml)	Saccharin, 95 mg	2143	(−)
	Drugs			
[0.3]	Phenacetin pill (average dose)	Phenacetin, 300 mg	1246	(2137)
[5.6]	Metronidazole (therapeutic dose)	Metronidazole, 2000 mg	(542)	506
[14]	Isoniazid pill (prophylactic dose)	Isoniazid, 300 mg	(150)	30
16*	Phenobarbital, one sleeping pill	Phenobarbital, 60 mg	(+)	5.5
17*	Clofibrate (average daily dose)	Clofibrate, 2000 mg	169	(?)
	Occupational exposure			
5.8	Formaldehyde: Workers' average daily intake	Formaldehyde, 6.1 mg	1.5	(44)
140	EDB: Workers' daily intake (high exposure)	Ethylene dibromide, 150 mg	1.5	(5.1)

*Asterisks indicate HERP from carcinogens thought to be nongenotoxic.

Potency of carcinogens: A number in parentheses indicates a TD_{50} value not used in the calculation of the human exposure–rodent potency (HERP) index because it is the less sensitive species. Symbols: −, negative in cancer test; +, positive for carcinogenicity in test(s) not suitable for calculating a TD_{50}; ?, not adequately tested for carcinogenicity. TD_{50} values shown are averages calculated by taking the harmonic mean of the TD_{50}s of the positive tests in that species from the Carcinogenic Potency Database. Results are similar if the lowest TD_{50} value (most potent) is used instead. For each test, the target site with the lowest TD_{50} value has been used. The average TD_{50} has been calculated separately for rats and mice, and the more sensitive species is used for calculating the possible hazard. The database, with references to the source of the cancer tests, is complete for tests published through 1984 and for the National Toxicology Program bioassays through June 1986. We have not indicated the route of exposure or target sites or other particulars of each test, although these are reported in the database. Daily human exposure: We have tried to use average or reasonable daily intakes to facilitate comparisons. In several cases, such as contaminated well water or factory exposure to ethylene dibromide (EDB), average intake is difficult to determine, and we give the value for the worst found. The calculations assume a daily dose for a lifetime; where drugs are normally taken for only a short period, we have bracketed the HERP value. For inhalation exposures, we assume an inhalation of 9,600 L/8 h for the workplace and 10,800 L/14 h for indoor air at home. Possible hazard: The amount of rodent carcinogen indicated under carcinogen dose is divided by 70 kg to give milligrams per kilogram of human exposure, and this human dose is given as the percentage of the TD_{50} dose in the rodent (in milligrams per kilogram) to calculate the HERP index. (Reprinted with permission from ref. 1. Copyright 1987 American Association for the Advancement of Science.)

An Alternative to Using Synthetic Pesticides is to Raise the Level of
Natural Plant Toxins by Breeding. It is not clear that the latter
approach, even where feasible, is preferable. One consequence of
disproportionate concern about tiny traces of synthetic pesticide
residues, such as ethylene dibromide (1), is that plant breeders are
developing highly insect-resistant plants, thus creating other risks.
Two recent cases are instructive. A major grower introduced a new
variety of highly insect-resistant celery into commerce. A flurry of
complaints to the Centers for Disease Control from all over the
country soon resulted, because people who handled the celery
developed a severe rash when they were subsequently exposed to
sunlight. Some detective work uncovered that the pest-resistant
celery contained 9000 ppb psoralens (light-activated mutagenic
carcinogens) instead of the level of 900 ppb psoralens present in
normal celery (29,30). It is unclear whether other natural
pesticides in the celery were increased as well.

Solanine and chaconine (the main natural alkaloids in potatoes)
are cholinesterase inhibitors and were widely introduced into the
human diet about 400 years ago with the dissemination of the potato
from the Andes. They can be detected in the blood of all potato
eaters. Total alkaloids are present in potatoes at a level of 15,000
μg per 200-g potato, which is only about a six-fold safety margin
from the toxic level for humans (1). Neither alkaloid has been
tested for carcinogenicity. By contrast, the pesticide malathion,
the main synthetic organophosphate cholinesterase inhibitor present
in our diet (17 μg/day), has been thoroughly tested and is not a
carcinogen in rodents. Plant breeders have produced an insect-
resistant potato; however, it had to be withdrawn from the market
because of its acute toxicity to humans, a consequence of higher
levels of solanine and chaconine.

There is a tendency for laymen to think of chemicals as being
only man-made, and to characterize them as toxic, as if every natural
chemical was not also toxic at some dose. Even a recent NRC/NAS
report (28) states: "Advances in classical plant breeding. . .offer
some promise for nonchemical pest control in the future. Nonchemical
approaches will be encouraged by tolerance revocations if more
profitable chemical controls are not available. . . ." The report
was particularly concerned with some pesticides used on tomatoes. Of
course, tomatine, one of the alkaloids in tomatoes, is a chemical
too, and was introduced from the new world 400 years ago. It has not
been tested in rodent cancer bioassays, is present at 36,000 μg/100 g
tomato, and is orders of magnitude closer to the toxic level than are
man-made pesticide residues found on tomatoes.

The idea that nature is benign and that evolution has allowed us
to cope with the toxic chemicals in the natural world (31) is not
compelling (20) for several reasons: (i) There is no reason to think
that natural selection should eliminate the hazard of carcinogenicity
of a plant toxin that causes cancer past the reproductive age, though
there could be selection for resistance to the acute effects of
particular carcinogens. For example, aflatoxin, a mold toxin that
presumably arose early in evolution, causes cancer in trout, rats,
mice, and monkeys, and probably people, though the species are not
equally sensitive (18,32). Many of the common metal salts are
carcinogens (such as lead, cadmium, beryllium, nickel, chromium,
selenium, and arsenic) despite their presence during all of

evolution. (ii) It is argued by some that humans, as opposed to rats or mice, may have developed resistance to each specific plant toxin or chemical in cooked food (31). This is unlikely, because, as discussed (1,21,22), both rodents and humans have developed many types of general defenses against the large amounts and enormous variety of toxic chemicals in plants (nature's pesticides). These defenses include the constant shedding of the surface layer of cells of the digestive system, the glutathione transferases for detoxifying alkylating agents, the active excretion of hydrophobic toxins out of liver or intestinal cells, numerous defenses against oxygen radicals, and DNA excision repair. The fact that defenses usually are general, rather than specific for each chemical, makes good evolutionary sense and is supported by various studies. Experimental evidence indicates that these general defenses are effective against both natural and synthetic compounds (33), since basic mechanisms of carcinogenesis are not unique to either. (iii) Most natural pesticides, like man-made pesticides, are relatively new to the modern diet, because most of our plant foods were brought to Europe within the last 500 years from the Americas, Africa, and Asia, and vice versa. (iv) The argument that plants contain anticarcinogens which protect us against plant carcinogens is irrelevant: plant antioxidants, the major known type of ingested anticarcinogen, do not distinguish whether oxidant carcinogens are synthetic or natural in origin, and thus help to protect us against both. (v) It has been argued that synthetic carcinogens can be synergistic with each other. However, this is also true of natural chemicals and is irrelevant to the argument that synthetic pesticide residues in food or water pollution appear to be a trivial increment over the background of natural carcinogens.

TCDD (Dioxin) Compared to Alcohol and Broccoli. Common sense suggests that a chemical pollutant should not be treated as a significant hazard if its possible hazard level is far below that of common food items. TCDD is a substance of great public concern because it is an extremely potent carcinogen and teratogen in rodents, yet the doses humans are exposed to are very low relative to the effective level in rodents. It is analyzed in some detail below to illustrate this point. TCDD can be compared to alcohol, as an example. Alcohol is an extremely weak carcinogen and teratogen, yet the doses humans are exposed to are very high relative to the effective dose in rodents (or humans). Indeed, alcoholic beverages are the most important known human teratogen, and the effective (5 drinks/day) dose level of alcohol in humans in mg/kg is similar to the level causing birth defects in mice. By contrast, there is no convincing evidence that TCDD is carcinogenic or teratogenic in man, though it is in rodents. If one compares the teratogenic potential of TCDD to that of alcohol for causing birth defects, after adjusting for their potency in rodents (1), then a daily consumption of the EPA "reference dose" (formerly "acceptable dose limit") of TCDD, 6 fg/kg/day, is equivalent in teratogenic potential to the amount of alcohol ingested daily from 1/3,000,000 of a beer, the equivalent of drinking one beer (15 grams ethyl alcohol) over a period of 8,000 years. A daily slice of bread, or glass of orange juice, contains much more natural alcohol than this.
    Alcoholic beverages in man are clearly carcinogenic, though only one of several tests on ethyl alcohol in rats was positive (12,13).

This test should be replicated as confirmation that ethyl alcohol is the active ingredient, though the evidence for that is fairly strong (12). A comparison of the carcinogenic potential of TCDD with that of alcohol, adjusting for potency in rodents, shows the equivalence for the TCDD reference dose of 6 fg/kg/day is 1 beer every 345 years. Since the average consumption of alcohol in the U.S. is equivalent to more than one beer per day per capita, the great concern over TCDD at levels in the range of the reference dose seems unreasonable.

The assumption of a worst-case linear dose response, often used for carcinogens, is not plausible for TCDD, yet extrapolations to man using those assumptions have generated great concern. TCDD binds to a receptor in mammalian cells, the Ah receptor, and the evidence suggests strongly that all of the TCDD effects are through this binding (34). Moreover, there is a wide variety of natural substances that bind to the Ah receptor, and as far as they have been examined they have all of the properties of TCDD. A cooked steak contains polycyclic hydrocarbons, which bind to the Ah receptor and mimic TCDD. In addition, our diet contains a variety of flavones and other substances from plants, which bind to the Ah receptor. The most interesting of such substances is indole carbinol (IC), which is present in large amounts in broccoli (500 mg/kg), cabbage, cauliflower, and other members of the Brassica family (35). IC induces the same set of enzymes as TCDD (36). When given before aflatoxin or other carcinogens, it protects against carcinogenesis, as does TCDD (37). However, when it is given after aflatoxin or other carcinogens, it is a strong promoter of carcinogenesis, as is TCDD (38). This stimulation of carcinogenesis has also been shown for cabbage itself (39). When IC is exposed to acid pH (equivalent to that of the stomach), it is converted to a series of dimers and trimers that are similar to TCDD in size and shape. These bind to the Ah receptor, and induce the set of TCDD-inducible enzymes, thus mimicking TCDD (36) (Bradfield, C.; Bjeldanes, L., University of California, Berkeley, personal communication, 1988). The 360 fg of TCDD/day EPA "reference dose" should be compared with 50 mg of IC per 100 g of broccoli, (one portion). Since the affinity of the indole derivatives in binding to the Ah receptor is less by a factor of about 8,000, the broccoli portion might be roughly 20 million times the possible hazard. Though these IC derivatives appear to be much more of a possible hazard than TCDD, it is not clear whether at these low doses either is any hazard.

Another study (40) also shows that when sunlight oxidizes tryptophan, a normal amino acid, it converts it to a variety of indoles (similar to the broccoli IC dimers), which bind to the Ah receptor and mimic the action of TCDD. It seems likely that many more of these "natural dioxins" will be discovered in the future.

Teratogens. The trace amounts of man-made pesticides found in polluted wells or on food should be a negligible cause of birth defects, when compared to the level of the background of known teratogens such as alcohol. Most agents causing birth defects would also be expected to be harmless at low doses. Important risk factors for birth defects in humans include: age of mother, alcohol, smoking, and rubella virus.

Cooking Food. The cooking of food generates a variety of mutagens and carcinogens. The total amount of browned and burnt material eaten in a typical day is at least several hundred times more than that inhaled from severe outdoor air pollution (22). Nine heterocyclic amines, isolated on the basis of their mutagenicity from proteins or amino acids that were heated in ways that reproduce cooking methods, have now been tested; all have been shown to be potent carcinogens in rodents (41,42). Many others are still being isolated and characterized (41,42). Three mutagenic nitropyrenes present in diesel exhaust have now been shown to be carcinogens (43), but the intake of these carcinogenic nitropyrenes has been estimated to be much higher from grilled chicken than from air pollution (41,42,44).

Gas flames generate NO_2, which can form both the carcinogenic nitropyrenes (13,15) and the potently carcinogenic nitrosamines in food cooked in gas ovens, such as fish. It seems likely that food cooked in gas ovens may be a major source of dietary nitrosamines and nitropyrenes.

Occupational Exposures. Pesticides could be of possible significance to chemical workers or applicators. Occupational exposures to chemicals can be high and can often be significant (1,45). The potential carcinogenic hazards to U.S. workers has been ranked using the PERP index (analogous to the HERP index (1) except that U.S. OSHA Permitted Exposure Levels replace actual exposures) (45). The PERP values differ by more than 100,000 fold. For 12 substances, the permitted levels for workers are greater than 10% of the rodent TD_{50} dose. Priority attention should be given to reduction of the allowable worker exposures that appear most hazardous in the PERP ranking.

We Cannot Extrapolate Risks Without Understanding Mechanisms of Carcinogenesis

Extrapolating Rodent Cancer Test Results to Humans. It is prudent to assume that if a chemical is a carcinogen in rats and mice at the maximum tolerated dose (MTD), it is also likely to be a carcinogen in humans at the MTD. However, until we understand more about mechanisms, we cannot reliably predict risk to humans at low doses, often hundreds of thousands of times below the dose where an effect is observed in rodents. Thus, quantitative risk assessment is currently not scientifically possible (1,17,20).

Carcinogenesis Mechanisms and the Dose-Response Curve. The field of mechanisms in carcinogenesis is developing rapidly and is essential for rational risk assessment. Both mutations and cell proliferation (i.e., promotion) are involved in carcinogenesis (1,46). There is an enormous spontaneous rate of damage to DNA from endogenous oxidants and methylation, which we have discussed in relation to cancer and aging (47-50). There is also a basal spontaneous rate for cell proliferation (7,14). Thus, increasing either mutation or cell proliferation should frequently be carcinogenic. Additional complications are that several mutations appear necessary for carcinogenesis, and there are many layers of defense against carcinogens. These considerations suggest a sub-linear dose-response relation, which is consistent with both the animal and human data

(51), and that multiplicative interactions will be common in human cancer causation. Most importantly, administering chemicals in cancer tests at the MTD might commonly cause cell proliferation (1,46,52,53). If a chemical is non-mutagenic and its carcinogenicity is due to its toxicity causing cell proliferation resulting from near-toxic doses, one might commonly expect a threshold (1,46,52).

The fact that high doses of a chemical cause tumors doesn't necessarily mean that small doses will. Most chemicals may, in fact, be harmless at low levels. A list of carcinogens isn't enough. The main rule in toxicology is that the "dose makes the poison": at some level, every chemical becomes toxic, but there are safe levels below that. In dealing with carcinogens, a scientific consensus evolved in the 1970s that we should treat carcinogens differently, that we should assume that even low doses could possibly cause some harm, even though we don't have the methods to measure effects at low levels. This idea evolved because most carcinogens appeared to be mutagens, agents that damage the DNA. The precedent of radiation, which is both a mutagen and carcinogen, gave credence to thinking that there possibly could be effects of chemicals even at low doses.

The idea that most of the classical carcinogens were mutagens damaging DNA (about 90% in our studies) (54,55) and work on oncogenes (56) reinforced the mutagen-carcinogen connection. However, in recent years there has been a change in the picture. About half of all chemicals tested in animals are carcinogens, and only about half of these appear to be mutagenic. It is now standard in cancer tests to be rigorous about giving the maximum tolerated dose of the chemical for the lifetime of the animal, and this factor, or the possibility that classical carcinogens comprised a special group, may account for the change in percentages. It seems quite reasonable that non-mutagens cause cancer because dosing at the MTD accelerates the promotional step of carcinogenesis (57,58).

Promotion, or cell proliferation, can also be accelerated by viruses, such as the human carcinogenic viruses hepatitis B, a major cause of liver cancer in the world (16), or human papilloma virus 16 (HPV16), a contributor to cancer of the cervix (15). Both cause chronic cell killing and consequent cell proliferation. Promotion can also be caused by hormones, which cause cell proliferation. Hormones appear to be major risk factors for certain human cancers such as breast cancer (14). The promotional step of cancer causation can also be accelerated by chemicals. Alcohol, for example, causes cirrhosis of the liver leading to cancer. The classical chemical promoters such as phenobarbital and tetradecanoyl phorbol acetate would be expected to be, and are in fact, carcinogens in animals when tested thoroughly (58). There is increasing evidence to show that low doses of promoters are not active (46,52). It seems likely, therefore, that a high percentage of all of the chemicals in the world, both man-made and natural, will be classified as carcinogens, but most of these may be acting as promoters and therefore may not be of interest at doses much below the toxic dose (1).

Epidemiological Fallacies: Storks Bring Babies and Pollution Causes Cancer and Birth Defects

The number of storks in Europe has been decreasing for decades. At the same time, the European birth rate has also been decreasing. We would be foolish to accept this high correlation (59) as evidence that storks bring babies. The science of epidemiology tries to sort out from the myriad chance correlations those that are meaningful and involve cause and effect. However, it is not easy to obtain convincing evidence by epidemiological methods because of inherent methodological difficulties (24). There are many sources of bias in observational data and chance variation is also an important factor. For example, because there are so many different types of cancer or birth defects, by chance alone one might expect some of them to occur at a high frequency in a small community. Toxicology provides evidence to help decide whether an observed correlation might be causal or accidental.

There is no convincing evidence from epidemiology or toxicology that pollution is a significant source of birth defects and cancer. For example, the epidemiological studies on Love Canal, dioxin in Agent Orange (60,61), Contra Costa County refineries (62,63), Silicon Valley (64), Woburn (1,21), or DDT provide no convincing evidence that in any of these well publicized exposures pollution was the cause of human harm. Even in Love Canal, where people were living next to a toxic waste dump, the epidemiological evidence for an effect on public health is equivocal. Analysis of the toxicology data on many of these cases suggests that the amounts of the chemicals involved were much too low relative to the background of natural and traditional carcinogens to be credible sources of increased cancer to humans (1). A comparative analysis of teratogens using a HERP-type index expressing the human exposure level as a percentage of the dose level effective in rodents would be of interest, but it has not been done in a systematic way.

Environmental exposure to ethylene dibromide and other pesticide pollutants is thousands of times lower than the exposure to these same agents in the workplace (1,45) (Table 1). Thus, if parts per billion of these pollutants were causing cancer or birth defects, one might expect to see an effect in the workplace. The studies on these chemicals so far do not provide any evidence for a causal association (32), though epidemiological studies are inherently insensitive. Historically, cases of cancer due to workplace exposure resulted mainly from exposures to chemicals at very high levels. For example, the EDB levels that workers were allowed to be exposed to were shockingly high (Table 1). (I testified in California in 1981 that our calculations showed that the workers were allowed to breathe in a dose higher than the dose that gave half of the rats cancer.) California lowered the permissible worker exposure over 100-fold. Despite the fact that the epidemiology on EDB in highly exposed workers does not show any significant effect, the uncertainties of our knowledge make it important to have strict rules about workers because they can be exposed to extremely high doses.

Technology is Not Doing Us In

Modern technologies are almost always replacing older, more hazardous technologies. The reason that billions of pounds of the solvents TCE (one of the most important industrial non-flammable solvents) and PCE (the main dry-cleaning solvent in the U.S.) are used is because of their low toxicity and the fact that they are not flammable. Is it advisable to go back to the age when industry or dry cleaners used flammable solvents and were frequently going up in flames? Eliminating a carcinogen may not always be a good idea. For example, ethylene dibromide (EDB), the main fumigant in the U.S. before it was banned, was present in trivial amounts in our food: the average daily intake was about one-tenth of the possible carcinogenic hazard of the aflatoxin in the average peanut butter sandwich, a trivial risk in itself (1) (Table 1). Elimination of fumigation results in insect infestation and subsequent contamination of grain by carcinogen-producing molds. This might result in a regression in public health, not an advance, and would also greatly increase costs. The coming alternatives, such as irradiating food, could be more hazardous than EDB, as well as more expensive. Similarly, modern pesticides replaced more hazardous substances such as lead arsenate, one of the major pesticides before the modern era. Lead and arsenic are both natural, highly toxic, and carcinogenic. Pesticides have increased crop yields and brought down the price of foods, a major public health advance.

Every living thing and every industry "pollutes" to some extent. How much does society wish to spend to get rid of the last part per billion of TCE out of wells in Silicon Valley, or PCE from dry-cleaning plants? We are currently spending enormous amounts of money trying to eliminate lower and lower amounts of pollution; one estimate is about 80 billion dollars annually (20); for comparison, the amount spent on all basic scientific research is $9 billion. The fact that scientists have developed methods to measure parts per billion (one part per billion is one person in all of China) of carcinogens and are developing methods to measure parts per trillion does not mean that significant pollution is increasing, or that the pollution found is a cause of human harm.

Conclusion

Everyone knows that spending all of one's effort on trivia without focusing on important problems is counterproductive. If we divert too much of our attention to traces of pollution and away from important public health concerns such as smoking (400,000 deaths per year), alcohol (100,000 deaths per year), eating unbalanced diets (e.g., too much saturated fat and cholesterol), AIDS, radioactive radon coming up from the soil into our homes, and high-dose occupational exposure, we do not improve public health, and the important hazards are lost in the confusion. It is the inexorable progress of modern technology and scientific research that will continue to provide the knowledge that will result in steady progress to decrease cancer and birth defects and lengthen lifespan.

Acknowledgments

I wish to thank Lois Gold and David Freedman for helpful discussion and criticisms. This work was supported by NCI Outstanding Investigator Grant CA39910 and by NIEHS Center Grant ES01896. This paper has been adapted from ref. 65.

Literature Cited

1. Ames, B. N.; Magaw, R.; Gold, L. S. Science 1987, 236, 271-280.

2. Gold, L. S.; Bernstein, L.; Magaw, R.; Slone, T.H. Environ. Health Perspect. 1989, in press.
3. Gold, L. S.; Sawyer, C. B.; Magaw, R.; Backman, G. M.; de Veciana, M.; Levinson, R.; Hooper, N. K.; Havender, W. R.; Bernstein, L.; Peto, R.; Pike, M. C.; Ames, B. N. Environ. Health Perspect. 1984, 58, 9-319.
4. Gold, L. S.; Slone, T. H.; Backman, G.; Magaw, R.; Da Costa, M.; Ames, B. N. Environ. Health Perspect. 1987, 74, 237-329.
5. Innes, J. R. M.; Ulland, B. M.; Valerio, M. G.; Petrucelli, L.; Fishbein, L.; Hart, E. R.; Pallotta, A. J.; Bates, R. R.; Falk, H. L.; Gart, J. J.; Klein, M.; Mitchell, I.; Peters, J. J. Natl. Cancer. Inst. 1969, 42, 1101-1114.
6. Schardein, J. L.; Schwetz, B. A.; Kenal, M. F. Environ. Health Perspect. 1985, 61, 55-67.
7. Lipkin, M. Cancer Res. 1988, 48, 235-245.
8. Yang, C. S.; Newmark, H. L. CRC Crit. Rev. Oncol./Hematol. 1987, 7, 267-287.
9. Pence, B. C.; Buddingh, F. Carcinogenesis 1988, 9, 187-190.
10. Reddy, B. S.; Cohen, L. A.; Eds. Diet, Nutrition, and Cancer: A Critical Evaluation, Vols. I and II; CRC Press: Boca Raton, FL, 1986.
11. Joossens, J. V.; Hill, M. J.; Geboers, J.; Eds. Diet and Human Carcinogenesis; Elsevier Science Publishers B.V.: Amsterdam, 1986.
12. Ames, B. N. Review of Evidence for Alcohol-Related Carcinogenesis. Report for Proposition 65 Meeting, Sacramento, CA, December 11, 1987.
13. IARC Monographs on the Evaluation of Carcinogenic Risks to Humans: Alcohol Drinking, Vol. 44, International Agency for Research on Cancer: Lyon, 1988, in press.
14. Henderson, B. E.; Ross, R.; Bernstein, L. Cancer Res. 1988, 48, 246-253.
15. Peto, R.; zur Hausen, H.; Eds. Banbury Report 21. Viral Etiology of Cervical Cancer; Cold Spring Harbor Laboratory: Cold Spring Harbor, NY, 1986.
16. Yeh, F.-S.; Mo, C.-C.; Luo, S.; Henderson, B. E.; Tong, M. J.; Yu, M. C. Cancer Res. 1985, 45, 872-873.
17. Ames, B. N.; Magaw, R.; Gold, L. S. Science 1987, 237, 235.
18. Ames, B. N.; Magaw, R.; Gold, L. S. Science 1987, 237, 1283-1284.
19. Ames, B. N.; Gold, L. S.; Magaw, R. Science 1987, 237, 1399-1400.
20. Ames, B. N.; Gold, L. S. Science 1987, 238, 1634.
21. Ames, B. N.; Gold, L. S. Science 1988, 240, 1045-1047.
22. Ames, B. N. Science 1983, 221, 1256-1264.
23. Doll, R.; Peto, R. The Causes of Cancer; Oxford University Press: Oxford, England, 1981.
24. Higginson, J. Cancer Res. 1988, 48, 1381-1389.
25. Peto, R. In Assessment of Risk from Low-Level Exposure to Radiation and Chemicals; Woodhead, A. D.; Shellabarger, C. J.; Pond, V.; Hollaender, A.; Eds.; Plenum: New York, 1985; pp. 3-16.
26. Harborne, J. B. In Natural Resistance of Plants to Pests. Roles of Allelochemicals (ACS Symposium 296); Green, M. B.; Hedin, P. A.; Eds.; American Chemical Society: Washington, D.C., 1986; Ch. 3.

27. Gartrell, M. J.; Craun, J. C.; Podrebarac, D. S.; Gunderson, E. L. J. Assoc. Off. Anal. Chem. 1986, 69, 146.
28. National Research Council, Board on Agriculture. Regulating Pesticides in Food; National Academy Press: Washington, D.C., 1987.
29. Berkley, S. F.; Hightower, A. W.; Beier, R. C.; Fleming, D. W.; Brokopp, C. D.; Ivie, G. W.; Broome, C. V. Ann. Intern. Med. 1986, 105, 351-355.
30. Seligman, P. J.; Mathias, C. G. T.; O'Malley, M. A.; Beier, R. C.; Fehrs, L. J.; Serrill, W. S.; Halperin, W. E. Arch. Dermatol. 1987, 123, 1478-1482.
31. Davis, D. L. Science 1987, 238, 1633-1634.
32. IARC Monographs on the Evaluation of Carcinogenic Risks to Humans. Overall Evaluations of Carcinogenicity: An Updating of IARC Monographs Volumes 1-42 (Supplement 7); International Agency for Research on Cancer: Lyon, France, 1987.
33. Jakoby, W. B., Ed. Enzymatic Basis of Detoxification, Vols. I and II; Academic Press: New York, 1980.
34. Knutson, J. C.; Poland, A. Cell 1982, 30, 225-234.
35. Bradfield, C. A.; Bjeldanes, L. F. J. Agric. Food Chem. 1987, 35, 46-49.
36. Bradfield, C. A.; Bjeldanes, L. F. J. Toxicol. Environ. Health 1987, 21, 311-323.
37. Dashwood, R. H.; Arbogast, D. N.; Fong, A. T.; Hendricks, J. D.; Bailey, G. S. Carcinogenesis 1988, 9, 427-432.
38. Bailey, G. S.; Hendricks, J. D.; Shelton, D. W.; Nixon, J. E.; Pawlowski, N. E. J. Natl. Cancer. Inst. 1987, 78, 931-934.
39. Birt, D. F.; Pelling, J. C.; Pour, P. M.; Tibbels, M. G.; Schweickert, L.; Bresnick, E. Carcinogenesis 1987, 8, 913-917.
40. Rannug, A.; Rannug, U.; Rosenkranz, H. S.; Winqvist, L.; Westerholm, R.; Agurell, E.; Grafstrom, A.-K. J. Biol. Chem. 1987, 262, 15422-15427.
41. Sugimura, T.; Sato, S.; Ohgaki, H.; Takayama, S.; Nagao, M.; Wakabayashi, K. In Genetic Toxicology of the Diet; Knudsen, I., Ed.; Alan R. Liss: New York, 1986; pp. 85-107.
42. Sugimura, T. Science 1986, 233, 312-318.
43. Ohgaki, H.; Hasegawa, H.; Kato, T.; Negishi, C.; Sato, S.; Sugimura, T. Cancer Lett. 1985, 25, 239-245.
44. Kinouchi, T.; Tsutsui, H.; Ohnishi, Y. Mutation Res. 1986, 171, 105-113.
45. Gold, L. S.; Backman, G. M.; Hooper, N. K.; Peto, R. Environ. Health Perspect. 1987, 76, 211-219.
46. Pitot, H. C.; Goldsworthy, T. L.; Moran, S.; Kennan, W.; Glauert, H. P.; Maronpot, R. R.; Campbell, H. A. Carcinogenesis 1987, 8, 1491-1499.
47. Park, J.-W.; Ames, B. N. Proc. Natl. Acad. Sci. USA 1988, 85, 7467-7470 and 85, 9508.
48. Ames, B. N. In Medical, Biochemical and Chemical Aspects of Free Radicals 1988; Yoshikawa, T., Ed.; Elsevier Science Publishers, B.V.: Amsterdam, 1989, in press.
49. Adelman, R.; Saul, R. L.; Ames, B. N. Proc. Natl. Acad. Sci. USA 1988, 85, 2706-2708.
50. Richter, C.; Park, J.-W.; Ames, B. N. Proc. Natl. Acad. Sci. USA 1988, 85, 6465-6467.

51. 1987 Annual Cancer Statistics Review Including Cancer Trends: 1950-1985 (NIH Publication No. 88-2789); National Institutes of Health: Bethesda, MD, 1988.
52. Farber, E. Environ. Health Perspect. 1987, 75, 65-70.
53. Swenberg, J. A.; Richardson, F. C.; Boucheron, J. A.; Deal, F. H.; Belinsky, S. A.; Charbonneau, M.; Short, B. G. Environ. Health Perspect. 1987, 76, 57-63.
54. McCann, J.; Choi, E.; Yamasaki, E.; Ames, B. N. Proc. Natl. Acad. Sci. USA 1975, 72, 5135-5139.
55. McCann, J.; Ames, B. N. Proc. Natl. Acad. Sci. USA 1976, 73, 950-954.
56. Stowers, S. J.; Maronpot, R. R.; Reynolds, S. H.; Anderson, M. W. Environ. Health Perspect. 1987, 75, 81-86.
57. Butterworth, B. E.; Slaga, T. J.; Eds. Banbury Report 25. Nongenotoxic Mechanisms in Carcinogenesis; Cold Spring Harbor Laboratory: Cold Spring Harbor, NY, 1987.
58. Iversen, O. H., Ed. Theories of Carcinogenesis; Hemisphere: Washington, D.C., 1988.
59. Sies, H. Nature 1988, 332, 495.
60. Lathrop, G. D.; Machado, S. G.; Karrison, P. G.; Grubbs, W. D.; Thomas, W. F.; Wolfe, W. H.; Michalek, J. E.; Miner, J. C.; Peterson, M. R.; Ogerskok, R. W. Air Force Health Study: Epidemiologic Investigation of Health Effects in Air Force Personnel Following Exposure to Herbicides. First Follow-Up Examination Results; U.S. Air Force: Brooks Air Force Base, TX, 1987.
61. Gough, M. Dioxin, Agent Orange: The Facts; Plenum Press: New York, 1986.
62. Austin, D. F.; Nelson, V.; Swain, B.; Johnson, L.; Lum, S.; Flessel, P. Epidemiological Study of the Incidence of Cancer as Related to Industrial Emissions in Contra Costa County, California (NTIS Publication No. PB84-199785); U.S. Government Printing Office: Washington, D.C., 1984.
63. Smith, A. H.; Waller, K. Air Pollution and Cancer Incidence in Contra Costa County: Review and Recommendations. A Report Prepared for the Contra Costa County Department of Health Services; 1985.
64. California Department of Health Services, Epidemiological Studies and Services Section, Pregnancy Outcome in Santa Clara County, 1980-1985; California Department of Health Services: Berkeley, CA, 1988.
65. Ames, B. N. In Important Advances in Oncology 1989; DeVita, Jr., V. T.; Hellman, S.; Rosenberg, S. A.; Eds.; J. B. Lippincott: Philadelphia, 1989; pp. 237-247.

RECEIVED July 7, 1989

Author Index

Affiliation Index

Subject Index

Production: Elizabeth Ryan Harder and Paula M. Bérard
Indexing: Deborah H. Steiner
Acquisition: Cheryl Shanks

Elements typeset by Hot Type Ltd., Washington, DC
Printed and bound by Maple Press, York, PA

Paper meets minimum requirements of American National Standard
for Information Sciences—Permanence of Paper for Printed Library
Materials, ANSI Z39.48–1984   ∞

Other ACS Books

Chemical Structure Software for Personal Computers
Edited by Daniel E. Meyer, Wendy A. Warr, and Richard A. Love
ACS Professional Reference Book; 107 pp;
clothbound, ISBN 0–8412–1538–3; paperback, ISBN 0–8412–1539–1

Personal Computers for Scientists: A Byte at a Time
By Glenn I. Ouchi
276 pp; clothbound, ISBN 0–8412–1000–4; paperback, ISBN 0–8412–1001–2

Biotechnology and Materials Science: Chemistry for the Future
Edited by Mary L. Good
160 pp; clothbound, ISBN 0–8412–1472–7; paperback, ISBN 0–8412–1473–5

Polymeric Materials: Chemistry for the Future
By Joseph Alper and Gordon L. Nelson
110 pp; clothbound, ISBN 0–8412–1622–3; paperback, ISBN 0–8412–1613–4

The Language of Biotechnology: A Dictionary of Terms
By John M. Walker and Michael Cox
ACS Professional Reference Book; 256 pp;
clothbound, ISBN 0–8412–1489–1; paperback, ISBN 0–8412–1490–5

Cancer: The Outlaw Cell, Second Edition
Edited by Richard E. LaFond
274 pp; clothbound, ISBN 0–8412–1419–0; paperback, ISBN 0–8412–1420–4

Practical Statistics for the Physical Sciences
By Larry L. Havlicek
ACS Professional Reference Book; 198 pp; clothbound; ISBN 0–8412–1453–0

The Basics of Technical Communicating
By B. Edward Cain
ACS Professional Reference Book; 198 pp;
clothbound, ISBN 0–8412–1451–4; paperback, ISBN 0–8412–1452–2

The ACS Style Guide: A Manual for Authors and Editors
Edited by Janet S. Dodd
264 pp; clothbound, ISBN 0–8412–0917–0; paperback, ISBN 0–8412–0943–X

Chemistry and Crime: From Sherlock Holmes to Today's Courtroom
Edited by Samuel M. Gerber
135 pp; clothbound, ISBN 0–8412–0784–4; paperback, ISBN 0–8412–0785–2

For further information and a free catalog of ACS books, contact:
American Chemical Society
Distribution Office, Department 225
1155 16th Street, NW, Washington, DC 20036
Telephone 800–227–5558

7 Day Reserve